SITING IN EARTHQUAKE ZONES

Siting in earthquake zones

JOHN G.Z.Q. WANG
Professor Emeritus, CIGIS, Peoples Republic of China
P. Eng., Ontario, Canada
K. TIM LAW
Professor, Carleton University, Ottawa, Ontario, Canada

A.A. BALKEMA / ROTTERDAM / BROOKFIELD / 1994

On the cover:
Approach fill failure caused by the 1991 Talamanco Earthquake (magnitude = 7.4) in Costa Rica. The failure is due to loss of foundation support in the underlying saturated loose sand which liquefied during the earthquake. In the background of the picture is the collapsed bridge over River Bananito. The collapse is again linked to soil liquefaction impairing the middle pier supporting both spans of the bridge (Photo courtesy of Professor D. Lau, Carleton University, Canada).

Published by
A.A. Balkema, P.O. Box 1675, 3000 BR Rotterdam, Netherlands
A.A. Balkema Publishers, Old Post Road, Brookfield, VT 05036, USA

ISBN 90 5410 092 3

Printed in the Netherlands

Contents

Preface

Destructive earthquakes occur from time to time in different parts of the world. Their socio-economic impacts are particularly large in some developing nations located in seismically active regions, due to such factors as population density, living conditions and building standards. China, for instance, faces threats of disastrous earthquakes as seen in the 1976 Tangshan Earthquake ($M = 7.8$) where roughly 250,000 people were killed. Developed countries are also not immune to earthquake disasters. One of the most recent reminders is the 1989 Loma Prieta earthquake in California. The total cost of damage was estimated at $10 billion (US). Canada also faces earthquake hazards. Well known for their seismic vulnerability are the west coast of British Columbia, in particular the cities of Vancouver and Victoria and in the east, the Lower St. Lawrence Valley containing Quebec City.

One important way to mitigate possible earthquake damage is by a systematic method of siting for any given project. Siting in earthquake zone is essentially an estimation of the seismic effects likely to occur within or around a site, and of the nature of possible damages. Siting is conducted through a thorough study of the topographic, geomorphological and lithological features of the site by using the theory and practice of earthquake engineering and geotechnical engineering with reference to experience of historical events. The important role of siting in earthquake zones can be illustrated by considering a structurally sound building placed on a thick layer of soft soil with the foundation designed properly for the static condition. The stability of the building during an earthquake, however, depends on how the whole soil layer interact with the seismic waves and the building. Siting will provide the essential information for assessing such stability.

This book is aimed at providing guidance in conducting a proper siting in earthquake zones. It supplements existing Chinese, the United States and Canadian codes regarding methodology of siting based on recent advances in this field with special emphasis on practical application.

This book is composed of three main components: fundamental concepts, checklists for site investigation and detailed methods of site evaluation. The first chapter on fundamental concepts deals with the nature and applicability of the book, along with definitions on some controversial technical terms. The second chapter describes step-by-step site investigation procedures expressed in the form of checklists. Such an investigation is essential to site selection. Chapter 3 to 9 further illustrate the principle and methods included in the checklists. Each chapter deals with one of the following topics: earthquake ground motion, seismic hazard analysis for a site, evaluation of seismic parameters, seismic effects of fault and faulting, seismic liquefaction of soils, landslides and slope stability under seismic action and ground waving and its damaging effects.

A lot of experience cited in this book comes from China. The original references are therefore written in Chinese. Most of these references do not have an official English translation. To preserve the true identity of the Chinese references, they are separately listed in Chinese and are referred to by numbers in square brackets in the text.

This book would not have been completed without the support and contribution of some organizations and individuals. The book is an adaption of a report on a project entitled 'Development of Guidelines for Siting in Earthquake Zones'. The project was largely funded by the International Development Research Centre, Canada. Data on laboratory testing and in situ testing within the project have been carried out with the support of the Institute for Research in Construction, National Research Council of Canada, Ontario Hydro and Comprehensive Institute of Geotechnical Investigation and Surveying (CIGIS), Ministry of Construction, China. Dr. C.F. Lee of Ontario Hydro was instrumental in initiating the project. Mr. R.X. Zhang, Mr. J.M. Wang and Mr. M. Wang, all of CIGIS, participated in the compilation of information and in writing a parallel report in Chinese. Mr. J.M. Wang was also responsible for typing and the layout of the entire manuscript and for reviewing Chapter 4. His patience and painstaking work is fully appreciated.

J.G.Z.Q. Wang & K.T. Law
Ottawa, Ont., Canada
July, 1993

CHAPTER 1

Fundamentals

1.1 SITING IN VIEW OF EARTHQUAKE RISK AND HAZARDS

Earthquake is a natural hazard that occurs randomly in seismic areas all over the world. As a result, people and their properties, the land and all the facilities in such areas are facing threats of earthquake disasters.

Numerous historical events show that within a strong earthquake zone, the seismic damages vary from one site to another depending on the geotechnical and geological conditions of the site and the structural integrity of the facilities. By studying these conditions and following a systematic method of siting, it is possible to reduce the potential hazards caused by strong earthquakes.

The scope of this book deals with siting with regard to identifying and understanding the effects of the geotechnical and geological conditions prevalent at a site. The purpose of siting is therefore directed towards a safe and economical process for city planning, general conceptual design, seismic provisions for various structures and treatment of foundation soils, etc.

For sites with available information on their characteristics, a choice can be made based on which one is the least vulnerable to seismic damage. For virgin lands without any existing information on site characteristics, siting is essential to ascertain the suitability of the site for the planned project. For existing built-up areas, on the other hand, siting can provide information to evaluate the seismic hazards of a possible future event. Hence, emergency preparedness plans and aseismic measures to strengthen structures and to improve foundation soils can be implemented in advance. In general, siting in earthquake zones is an indispensable component of earthquake hazard mitigation.

The principle of siting described in this book is based on the seismo-geotechnical response of foundation soils subject to local site conditions and geological settings. While conventional site investigation methods may provide much important relevant information, they are covered in many available textbooks and are therefore not included in this book. As well, the readers of this book are expected to be familiar with the basics of earthquake engineering and geotechnical engineering. Basic theories and concepts are therefore either briefly discussed or omitted.

Siting in earthquake zones is a rather empirical approach to predict the seismic effects and impact on a site in the light of past experience and aided by theoretical analysis. Within this process, the evidence of past earthquake damage and its recurrence nature are the most important guides for assessing future event and its potential damage.

1.2 FUNDAMENTAL CONCEPTS

Due to existing divergent opinions and some confusions in earthquake engineering, it is necessary to discuss and define a number of conceptual terms in order to maintain clarity and consistency in integrating theory with practice.

1.2.1 *Microzonation*

Microzonation is the subdivision of a seismic zone into smaller zones (microzones) according to a certain criterion to facilitate the implementation of aseismic measures. One common criterion is expressed in terms of seismic intensity, with the result that different microzones will be associated with different hazards. Another criterion is in terms of seismic response spectrum. This microzoning will yield different ground motions for different microzones. Although these two approaches provide important supplementary information to the seismic zonation of larger areas, they fail to account for geological and topographical features that significantly affect the degree of damage due to an earthquake. For large construction projects, therefore, it is preferable to carry out microzonation by incorporating these features. This microzonation process is new and adopted in this book.

The site size should normally exceed the immediate confines of the building. It should include the geological and topographical environment that may decisively change the ground motion characteristics at the location of the building.

1.2.2 *Assessment of the site and foundation soils*

The term foundation soils refers to all the geological deposits, irrespective of their compactness, directly supporting and affected by the loading of the buildings. The seismic behaviour of a foundation soil is controlled by its dynamic behaviour and by the geological conditions of the site.

The assessment of site and foundation soil in this book is aimed at meeting the requirements of design and construction in earthquake zones. The method of assessment considers the general and specific characteristics of the site and the foundation soils. It analyzes the behaviour of the whole site and examines the details of each foundation soil component. The method, however, does not consider performance unrelated to earthquakes.

The term siting refers to the evaluation and prediction of effects and potential damages by future earthquake events on the site and the foundation soils. It is an important component of the overall seismic design.

1.2.3 *Uncertainty of earthquake – Importance of conceptual design*

Earthquake motion is the result of a complex interaction among a series of variables including the seismic source, the medium of wave propagation, the site conditions and foundation soil characteristics. Some of these variables are deterministic while the others are probabilistic. The resulting motion is therefore a complicated and fussy event. In addition, some seismic phenomena have dual nature, each with opposite effects. For example, surface faulting or soil liquefaction may either intensify or lessen the effects of damage, depending on conditions to be described in the following chapters. The net effects of such a complex interaction are often not adequately accounted for even by existing detailed technical design.

To offset this inadequacy, therefore, a conceptual design approach has been promoted since the last decade. This approach incorporates available accumulated empirical data and analysis methods. It assesses conceptually the relative importance of different variables and guides the correct application of the detailed technical design. Examples of this approach are given in Chapters 6 and 7.

1.2.4 *Recurrence of seismic damages*

Numerous historical events in China, Japan or the US indicate that ground failures and damages caused by earthquake shocks are clearly recurrent. Soil liquefaction and sand boiling, for example, occurred repeatedly at the same district, same site or even the same point of a foundation soil, during a series of strong earthquakes (Wang 1981; Wang et al. 1989). Moreover, tall chimneys and some high masonry structures suffer the same damage during successive strong earthquakes (Wang 1983). At a broader level, anomalies of high or low intensity zones have been located in the same area as in some historical events. The Marina district in the San Francisco Bay area is an example of anomaly of high intensity zone as observed in the past several earthquakes including the 1989 Loma Prieta Earthquake and the 1906 San Francisco Earthquake. Many anomaly spots of high intensity are consistently found in the same area in Beijing district, China, as evidenced by ancient buildings and structures sustaining the same destruction in successive events in these anomaly spots.

The recurrence of similar seismic damage in the same areas strongly indicates the role played by the geological, geo-morphological and lithological features of the site and by the behaviour of soil. These factors causing the recurrence are beyond the mechanism of seismic source, earthquake magnitude and epicentral distances. The reccurrence in the same area is therefore determined by the internal factors of the area. The role played by these factors are discussed in detail in later chapters.

The recurrence of seismic damage implies a certain regularity based upon which important lessons can be drawn. It enables engineers to predict the potential damage of given sites during future strong earthquakes. If the conditions of the site and the soil remain unchanged, one can almost be certain of what kinds of damage will reoccur. Therefore, investigation of the past recurring seismic damages in historical strong earthquake areas will yield realistic assessment of the effects of site and foundation soil. This can be accepted as a basic principle particularly for analyzing soil liquefaction in level ground and for surface faulting of tectonic origin.

1.2.5 *Continuity of seismic activities*

Seismic activities, including occurrence mechanism and ground motion transmission, are controlled by the geology of the earth's crust. For a given geological period, therefore, there is a certain continuity of earthquake activity. In terms of geological age, the past several thousand years can be considered as a single geological period. Within this period, seismic activities are considered recent and to display some continuity.

Engineers involved in seismic designs are concerned with the safe and economic construction of buildings and structures for the relatively short future, say, 100 years. Any seismic event within such short period may be considered a part of the recent seismic activities of the last several thousand years. Events that occurred within this last period are a prelude to the events to come in the near future. Such continuity of events provides engineers with some useful rules for assessing the impact of future events on their designs.

Seismic continuity provides important information in the time, space and intensity domains if sufficient historical data exist. In the time domain, seismic continuity forms the basis of probability and frequency of earthquake occurrence or recurrence relationship for different earthquake magnitudes. In the space domain, past records will provide engineers clues on the geographical distribution of earthquakes and their focal depths. In the intensity domain, seismic continuity may reveal the most probable magnitudes of fore-shocks, main shocks and after-shocks and their associated characteristics of the ground motion spectra. All these are basic parameters for seismic design.

1.3 MEANING OF SOME BASIC TERMS

A common terminology in earthquake engineering has been followed in this book. Some basic technical terms, however, do not have a unified definition. For maintaining clarity in this book, these basic terms are discussed here. Other specialized terms such as liquefaction and ground faulting will be discussed in the appropriate chapters.

Seismic/earthquake zone. Strictly speaking, earthquakes have occurred almost everywhere on the globe during its long history of existence. For engineering purposes, a meaningful definition for seismic or earthquake zone should be qualified by some measures of the time, location and intensity of future expected earthquake occurrences. In this book, earthquake zones are defined as zones expected to experience, in the coming 100 years, earthquake intensity over VI on the Modified Mercali Scale or a peak horizontal acceleration, $a_{max} \geq 0.08$ g, where g is the gravitational acceleration.

Earthquake action. This is a synonym of seismic force or seismic load. In fact, this type of force is produced indirectly through the ground motion. Again, seismic ground motion is understood as a fuzzy event which occurs randomly and unexpectedly with totally irregular scale of amplitude and duration of time. Therefore, such a seismic 'force' or 'load' is actually not a scaler quantity, and it is suggested to be referred to as seismic 'action'.

Epicentral area. This refers to the macroscopic area where maximum intensity occurs in an event. Sometimes it does not lie in the same location as the epicentre determined by graphical methods.

Earthquake intensity. The 12 grade intensity scale, used in China, Europe and the formerly Soviet Union, is basically similar to the Modified Mercali Scale (MMS) used in North America. However, the corresponding seismic acceleration given empirically differs significantly as shown in Table 1.3.1 This is due to the

Table 1.3.1. A comparison of seismic intensity versus acceleration specified in some countries/region.

China (I)		Japan (JMA)		North America (MMS)		Western Europe (MCS)		Eastern Europe (MSK)	
12 grades	a gal	8 grades	a gal	12 grades	a gal	12 grades	a gal	12 grades	a gal
I		0	< 0.8	I		I	0.25	I	
II		I	0.8-2.5	II	1.0	II	0.25-0.5	II	
III		II	2.5-8.0	III	1.6	III	0.5-1.0	III	
IV		II-III	2.5-8.0	IV	3.2	IV	1.0-2.5	IV	
V	31	III	8.0-25	V	10	V	2.5-5.0	V	12-25
VI	63	IV	25-80	VI	32	VI	5-10	VI	25-50
VII	125	IV-V	80-250	VII	79	VII	10-25	VII	50-100
VIII	250	V	80-250	VIII	200	VIII	25-50	VIII	100-200
IX	500	VI	250-400	IX	316	IX	50-100	IX	200-400
X	1000	VI	250-400	X	398	X	100-250	X	400-800
XI	> 1000	VII	> 400	XI		XI	250-500	XI	
XII		VII	> 400	XII		XII	500-1000	XII	

very limited quantity of earthquakes among different countries and thus these are mainly empirical estimates. Earthquake intensity is a comprehensive evaluation of the overall earthquake damages and the changes of the natural circumstances due to earthquake, hence it is being called 'macroscopic intensity'. On the other hand, it is used in some countries to serve as an important scale for predicting a future event, so it can be used as a parameter to identify earthquake zones.

Strong earthquake zones. The term used here is directed to those areas of earthquake intensity equal and over VII (12 grade scale). Due to such intensity, a certain amount of damage will certainly occur, so siting in such areas should take seismic aspects into consideration.

Far field/far source earthquake. The influence field of an earthquake normally extends outwards from the epicentre with a certain degradation of intensity. 'Far Field' denotes those areas with an intensity of two or more grades less than the epicentral area. Relatively speaking, such a seismic event is called a far source earthquake.

Near field/near source earthquake. Comparatively speaking, near field denotes those areas where the intensity difference is less than 2 with respect to the epicentral zone. Such a seismic event is called a near source earthquake.

CHAPTER 2

Major program and method of siting

2.1 ESSENTIALS OF SITING

Siting in earthquake zones is essentially a prediction of possible damages and/or seismic effects likely to occur within or around the scope of a site, through a thorough study of the topographic, geo-morphological and lithologic features of the site by using the theory and practice of earthquake engineering, geotechnical engineering and by referring to the experience of historical events.

This work can be implemented separately or combined with the seismic design in either the conceptual or the technical design phase.

2.2 DESIGN EARTHQUAKE

2.2.1 *Basic concept*

A 'design Earthquake' can be simply understood as an assumed earthquake which represents all the characteristics and behaviours of an earthquake likely to occur within a certain limit of time and location at the site. The use of the design earthquake is to provide a basis of seismic design for a certain project.

In the former Soviet Union, Easter Europe and China, two intensity terms have often been used: basic intensity and fortified intensity. Basic intensity is the design earthquake intensity (12 grades) used for ordinary structures. Fortified intensity is a modification of the basic intensity to account for the importance of the structure. For ordinary structures, the fortified intensity is equal to the basic intensity. For important structures, the fortified intensity is upgraded by one or two from the basic intensity depending on how important is the structure. For structures less important than the ordinary ones, the fortified intensity is downgraded to any intensity appropriate to the importance of the structure.

The given basic intensity or the fortified intensity officially specified by authority for a certain area is a large part of the design earthquake. However, it does not represent all the contents of the design earthquake.

2.2.2 *Major contents of design earthquake*

To determine the contents of a design earthquake, the following work is normally done wholly or partly according to the various conditions of sites:

1. Establish the seismicity characteristics of a local area or near field of interest at present and in the past, such as the characteristics of the frequency and amplitude of the major shock, the fore-shock and the after shock.

2. Based on regional seismo-geological background, assess the possible maximum magnitude likely to occur at the site. For this purpose, regional geological survey may be needed with emphasis on the seismo-geotectonic aspect.

3. Ascertain the most possible acceleration and the repeatable average peak acceleration in a future event. Therefore, the collection of historical seismic record is necessary, and by empirical method the characteristic acceleration specific to a certain area similar to the site of interest can be obtained. The spectrum of ground motion can also be estimated.

4. Determine the predominant period of ground motion. For preventing or decreasing damages due to resonance between the ground soil and the planned structure, the seismic designers should design the inherent period of vibration of the structure so that it is different from the natural period of the site (i.e. predominant period). The predominant period of ground motion of a site during earthquakes is normally consistent with or even equal to that obtained from the micro-tremor measurements under relatively quiet conditions on the same site. However, due to the lack of a thorough understanding of the mechanism of micro-tremor so far, it is suggested to use empirical methods to supplement the micro-tremor measurements. Table 2.2.1 (Kanai 1961) is a helpful reference of empirical relationship between the soil classification and the maximum or the average period of vibration. However, some practical experience shows that the predominant period of a site obtained from micro-tremor measurements may

Table 2.2.1. Classification of ground soils versus their period under vibration.

Classification no.	Classification of soils	$T_{0,\max}$ (s)	$T_{0,\text{av}}$ (s)
I	Bedrock or stiff sand gravel layers over a vast site surrounding the buildings and structures. Other places also embedded with tertiary or earlier deposits	0.3-0.70	0.05-0.2
II	Gravel and sand/stiff clay and loam inter-bedded over a vast site. Other places embedded with residual soils or sand-gravel alluvials of over 5 meter thickness	0.55-1.05	0.13-0.4
III	Soil in general	0.92-1.35	0.35-0.7
IV	(1) Humic soil, sea bed soft clay, and alluvial composed of similar soils with a thickness of over 30 meters. (2) Marsh, hydraulic fill with sea mud and the like deposited within 30 years	1.20-1.50	0.57-0.8

differ from that during an earthquake. Some records also show that the predominant period increases proportionally with magnitude. In general, the average value of predominant period is normally suggested in the investigation report.

5. Estimate the epicentral distance of the design earthquake. This is a parameter directly associated with the 'near field design earthquake' and the 'far field design earthquake'. In addition, seismic hazard analysis based on probability of exceedance of annual occurrence requires the minimum horizontal distance between the site and the causative fault (Section 4.3.2). Sometimes, this is also a parameter to determine whether or not a permanent surface faulting will occur.

6. Select the design Earthquake Spectrum. For those projects subject to special requirements for preventing earthquake damage and/or subject to unusual site conditions, one acceleration value may not be enough to reflect the dynamic behaviour of the site. It is necessary to choose the proper seismic record of a similar site in order to calculate the response spectra appropriate to meet the needs of the project. Normally, this is what the design spectrum can represent.

2.2.3 *Macroscopic approach to the background of design earthquake*

In order that the design earthquake fully reflects the characteristics of the future event in consideration, and that a reliable and reasonable aseismic design may be obtained, it is necessary to examine the background and parameters of the proposed design earthquake through a macroscopic study using the following checklist. More information for the rationale and for detailed description of procedure is given in later chapters.

1. *Earthquake parameters of historical events*

Check the following situation in the very region (site) or adjacent district (site) in a historical earthquake:

a) Is the magnitude (M) extremely inconsistent with the epicentral intensity (I_0)? If yes, further distinction is required. Does inconsistency of high intensity-low magnitude anomaly exist? Or does low intensity-high magnitude anomaly exist?

b) Is there any obvious connection between magnitude and length of surface faulting or any known deeply embedded major causative fault?

c) Is there any relationship between the focal depth and the size of felt area of earthquake? If any, indicate the details.

d) Is there any connection between the magnitude and the time duration? State the particular relationship, if any.

e) Is there any correlation among peak acceleration, velocity and epicentral intensity provided by the original record of a historical earthquake? Give the concrete correlation, if any.

2. *Surface faulting and source mechanism*
a) Is the surface faulting a tectonic one?
b) Did the surface faulting occur in the outcrop of bedrock?
c) Is the surface faulting an extension of deeply embedded causative fault up to the ground?
d) Did gravitational rupture occur?
e) Did the tectonic rupture occur on the ground? If any, visualize its mechanism including its strike, relative dislocation and size.
f) Is there any regular distribution of historical epicentres? Show evidence, if any.
g) Is there any periodical evidence of the happening of historical earthquakes? Sketch out on a map, if any.
h) Any correlation between the intensity distribution and the surface faulting (or causative fault)? Present schematically, if any.

2.3 INVESTIGATION ON THE SEISMIC EFFECT OF THE SITE AND GROUND SOIL IN HISTORICAL EVENTS

There are two main seismic effects of the site and ground soil:

Direct effects – the earthquake resisting behaviour of the site and ground soil is weak, resulting in a certain type of ground failure.

Indirect effects – the site and ground soil are strong enough to resist earthquake disturbance, but during their transmission of seismic waves and motions, they behave either as a destructive agent, or as a protective medium by providing attenuation and/or absorption of seismic energy.

2.3.1 *Direct seismic effects*

In general, the failure of the site and/or the ground soil under earthquake is apparently visible and commonly recognized. The main patterns of such a failure are as follows:

1. *Seismic liquefaction of soils*
This specific problem will be discussed later in Chapter 7. In this section, a special investigation checklist is given particularly for the assessment of macroscopic liquefaction which often occurs on level ground during a strong earthquake. The result can be used as a prediction of liquefaction potential of the site in a future event. The checklist is given in the following:

a) Is the basic intensity or the design earthquake intensity of the area in question in good agreement with the maximum intensity which took place in the past including all those greater than VI(I > VI) of the 12 grade scale?

b) Is there any saturated silty sand and/or fine sand existing 0.8 ~ 15.0 m below the level ground surface? If not, terminate investigation for seismic liquefaction.

c) Is the site situated in an unfavourable geo-morphological unit such as a river bend, delta deposit area, alluvial fan or oxbow lake with soil deposit?

d) Was the historical earthquake damage of a liquefied area relatively heavier than that of non-liquefied areas?

e) Is the site situated on a recently-deposited saturated soil or a buried ancient river valley?

f) Is there any evidence left on the site indicating past liquefaction experience, such as the drunk tree and drunk forest? Are there inclined old buildings due to uneven settlement induced by liquefaction? If any, state how it is distinguished from uneven settlement due to static loading.

g) Is there any trace of sand boiling, such as sand veins, sand seams in cohesive layers, etc.?

h) Is the suspected material loose?

2. *Seismic settlement*

Seismic settlement can be caused by different actions. The real cause should be identified at first, then relevant measures can be taken accordingly.

a) Did the settlement occur mainly in loose or non-saturated sand layers?

b) Did the settlement occur in a liquefied layer or silty soil?

c) Is there any trace of sand boiling in or near the settled ground?

d) Is there any connection between the magnitude of the settlement and the size of micro-geomorphological boundary?

3. *Collapsible settlement*

Collapsible settlement refers to the sudden settlement due to an earthquake-induced collapse of underground cavity or space which forms a discontinuous interface in the soil. This type of collapsible settlement differs in nature from the seismic settlement. The following should be carefully investigated to determine the correct settlement type.

a) Does the settlement have a clearly confined boundary?

b) Does the settlement have a relatively limited area in size?

c) Is there any underground space, such as a cavern, cavity, underground chamber, underground defrosted layer of permafrost or under-minings?

4. *Seismic landslide*

There are several reasons causing seismic landslides, which should be distinguished from one another:

a) Is the landslide due to soil liquefaction in a sloped ground or in a layer with an inclined bottom? (If yes, traces of liquefaction should be visible).

b) Is it simply due to a slope failure under seismic action?

c) Is it a result of the horizontal waving of the ground during the shaking of an earthquake? (If yes, there should be two areas moving in opposite directions).

5. *Destruction of foundation*
The confining action of the surrounding soil of a foundation will provide a more stable horizontal support and will therefore lead to less failure or destruction to a foundation structure itself rather than the superstructure on the ground. More importantly, the destruction or failure of a superstructure normally occurrs prior to that of the foundation structure, and once it has occurred, the seismic action (load) induced by the mass and through the rigidity of the superstructure will suddenly be decreased.

However, there have been few cases of foundation failure similar to the seismic damage on underground structures. Those failures should be analyzed according to the following checklist:

a) Is there any vertical displacement of the foundation structure? (e.g. any gaps beneath the bearing column, pier, caisson and/or ground slab? Any shear cracks on the ground beams? Any tensional cracks on the neck of the pile cap? Any diagonal tension cracks on the bearing wall due to uneven settlement of the foundation soil? etc.).

b) Is there any horizontal displacement of the foundation structure? (e.g. cracks on the basement wall, torsional movement of footings, horizontal shear on the pile cap, etc.).

2.3.2 *Indirect seismic effects*

Indirect seismic effects particularly refer to many of the influence of site and ground soil on the damage of the super-structures during the transmission and propagation of seismic waves. Although in most cases, the supporting material did not directly cause the superstructure to fail, the seismic waves transmitted through the material may be magnified either in amplitude or in time duration, as a consequence of the action of ground motion.

1. *Damages due to resonance*
Numerous case histories obtained from the investigation show that resonance damage used to be the most common pattern of earthquake hazards on the ground structures due to the invisible interaction between the site/ground soil and the structures. For further clarification, the following points should be stressed and checked:

a) Is the predominant period of the site and ground soil consistent with that of the damaged building (the period can be either measured or empirically estimated)?

b) Were tall and flexible structures (like masonry chimney stalk, etc.) in the far field destroyed more severely than those in the near field?

c) Were rigid structures (such as heavy rigid concrete frame, concrete block structure or boulder mortar structure, etc.) destroyed more severely on rocky ground than those on softer ground?

d) Were buildings/structures of different rigidity on the same site destroyed in different manners?

2. *Damaging effect of wave field*

A lot of ground facilities (such as railway, highway pavement, shallow-buried pipeline, bridge, canal, water trench, etc.) destroyed have not been due to seismic action (force) directly acting on them because their mass is relatively small and the overburden pressure is negligible or even zero. Seismic action exerted from the product of their mass and seismic acceleration is by no means large enough to destroy themselves even if the seismic acceleration is high.

In fact, all ground facilities are relatively fixed on the ground as they move together during earthquakes. Many traces reveal the evidence of ground waving which sometimes is of enlarged amplitude and in complicated (mainly due to standing wave, Love wave and Rayleigh wave) waving pattern (Chapter 9). As a result of such intolerable waving motion, damage is serious and overwhelming.

This type of seismic effect is actually caused by some special topographic, micro-geomorphological and soil layer conditions. Therefore such an effect is repeatable. If waving damage is confirmed as a historical evidence, it can be used as a basis for prediction and seismic design of reconstruction project on the same site or for those projects on similar site conditions. To this end, it is required to investigate the following:

a) Were any pipelines, railway, etc. on the ground damaged? If yes, describe the pattern of damage.

b) Was there any rhythmic or waving traces left on the linear structures? If yes, try to measure the geometry of those traces.

c) Did the waving trace occur at some specific localities (such as a river bend, river bed, river valley, scarp or sudden change of strata and lithology)?

3. *Amplification and filtering effects of soft ground on seismic wave*

In transmitting seismic wave from the bedrock on to the ground, the soil layer may amplify the amplitude of the wave and at the same time absorb part of the high frequency components. As a result, it prolongs the period and duration of earthquake shock.

The amplification and filtering effects are more obvious in soft soil and unfavourable to the structures of longer periods of vibration, because amplification may do much harm on structural ductility and may induce long periods and duration of vibration. Therefore, the following investigations are suggested:

a) In any historical event, was there any evidence showing ground motion ever been magnified? If any, specify the particular location and soil lithology.

b) Did the amplification indicate any orientation? If there are near/far field seismic records in X-Y-Z direction available, acquire the records.

c) Was there any filtering on the historical seismic records? If any, the near/far field record should be illustrated with emphasis on frequency characteristics.

d) Was there a big difference in damage between any two sites of similar epicentral distance demonstrating the influence of amplification and/or filtering due to different soil lithology? If any, real seismic record or actual measurement of the fundamental period is necessary.

2.4 KEY POINTS OF INVESTIGATING A PROPOSED SITE FOR CONSTRUCTION

In view of the complex nature of earthquake action and the large variety of media propagating that action (in geological and topographical sense), it is necessary to use the deductive method for assessment and evaluation of a district (site). Attention should be paid to searching for any regularity of the seismic effects from existing case histories and analyze the regularity in theoretical considerations.

2.4.1 *Major roles and programs*

1. Ascertain no severe earthquake hazard should occur, such as tectonic rupture, surface faulting, large area liquefaction, sliding, seismic settlement. These hazards may alter the overall stability of the site. As for the particular assessment of each individual problem, refer to the following chapters.

2. Finalize the design earthquake parameters (sometimes, the existing microzonation or relevant results can be used instead), including choosing input seismic data.

3. Measure relevant dynamic parameters both in the laboratory and in situ for analyzing interaction of ground soil and the superstructure, dynamic deformation, seismic settlement and dynamic response spectrum of the site, and for calculating dynamic earth pressure and hydro-dynamic pressure acting on retaining structures, earthdams and/or embankments when necessary.

The content of work to be done should be adjusted according to different requirements in various areas. If abundant past engineering experience is available the required work can be reduced.

2.4.2 *Major contents of work*

The major contents of work specified here are particularly pertinent to conceptual design for earthquake resistant projects. Those siting outside earthquake zones or those relating to non-seismic design are not included in this book.

The major items of work listed in the following are aimed at the development of major civil engineering projects (such as a new town, residential area, industrial district, large open mining with high open cut and deep excavation project, etc.) to be built in an earthquake zone with which the designer has little experience or information. As for other earthquake zones where experience or information are adequate, or for minor projects of less importance, the following work may be reduced.

1. *Surface geo-seismic survey*
Normally, supplementary survey is to be done on the basis of existing geological survey. Emphasis should be put on the following:

a) Can any dislocation be traced to or even visible in the quarternary deposit? If any suspicious surface rupture exists, sampling is necessary for dating and/or for micro-structure scanning to determine its nature.

b) Can any seismo-topographical feature be found, such as fractured valley, fault scarp, fault spring? If any, verify its age and behaviour.

c) Can any scarp or colluvial deposit due to earthquakes be found? If any, confirm size, original geological structure and its distribution.

d) Did any debris flow caused by earthquakes occur? If any, the structural characteristics of the debris is to be determined.

e) Can any drunk tree or inclined old structure suspectedly caused by earthquake be found? (Drunk trees and forest caused by earthquake is normally characterized by a large area of trees inclining at a certain angle and pointing to a certain direction as a group).

2. *Subsurface survey*
Exploration should be conducted down to a certain depth to meet the following requirements.

a) For calculating the dynamic response, exploration should reach a depth of 50 m for high buildings, tall structures or deep excavated foundations; 20 m for ordinary projects; or to bedrock or a hard soil layer (where shear wave velocity $V_s > 500$ m/s) encountered within those depths. A cross section of the subsurface conditions should be established.

Exploration for the most important projects should be extended down to 100 m or to bedrock or a hard soil layer if found before this depth.

b) For liquefaction assessment, exploration should go down to 15 m for a high building or heavy structures; or down to 10 m for common projects.

c) For detecting suspicious faults or fracture zones, exploration should be extended to 50 m for high buildings or heavy structures; or to 30 m for common projects.

d) For investigating suspicious seismic landslides and liquefaction induced slidings, exploration should be extended to 20 m below the toe of a large slide or to 10 m below the toe of a small size slide.

e) For detecting suspicious seismic settlements, exploration should be extended to a depth $H = L/(2 \tan 30°)$, where L is the maximum horizontal length of the settled area.

3. *Subsurface sampling and in-situ testing*

a) In principle, sampling and in-situ testing should reach the same depth as the above-mentioned exploration.

b) For dynamic response analysis, sampling and/or in-situ testing should be conducted at least once in every homogeneous soil layer. For those layers which are thinner than 10 cm, sampling and/ or in-situ testing for this purpose may be neglected (except for detecting seismic landslide).

c) For dynamic response analysis, in-situ tests are conducted mainly to determine the shear wave and longitudinal wave velocity (cross-hole method is preferable). Damping ratio and mass density are required in some cases.

d) For more specific purpose, e.g. liquefaction potential assessment, undisturbed sampling, the standard penetration test (SPT) and the cone penetration test (CPT), etc. are normally undertaken simultaneously.

e) For confirming the existence of past seismic landslides, specific sampling and/or in-situ testing should match exploration as well. Emphasis should be placed on identifying and measuring the behaviour of thin layer, especially the weak sandwich layers.

f) Shear wave velocity (V_s) and longitudinal wave velocity (V_p) are necessary for the following purposes;

– To calculate the seismic response of ground or the characteristic values of movement (displacement, velocity or acceleration) at the interface between layers;

– To calculate the Poisson's ratio, modulus of elasticity and shear modulus of ground soils;

– To establish relationship among soil density, compressibility and rigidity;

– To be used as criteria for evaluating liquefaction potential of soils based on statistical relationships.

g) Surface wave velocities – Rayleigh wave (V_r) and Love wave (V_l) can be measured by means of surface excitation on a concrete block with certain mass. Direction of impact excitation and the locations of pickups should be controlled according to the wave velocity to be measured. Normally, the uses of such measurements are as follows:

– To calculate the wave lengths of the surface waves;

– To evaluate effects of rhythmic damage (Chapter 9) on the ground in order to estimate the proper length of the structure.

4. *Laboratory testing*

Laboratory testing for siting in earthquake zones deals mainly with the dynamic properties and behaviours of soil sampled from the site of interest. The following methods are most commonly used [13]*:

*Number in square brackets refers to reference number in the Chinese reference list.

a) The dynamic triaxial test is a relatively versatile method. It can provide the basic dynamic stress-strain relationship of the soil, from which the dynamic modulus, dynamic damping ratio and dynamic strength can be derived. Liquefaction potential of soil under specified stress conditions and frequency can also be evaluated. However, low stress ratio and disturbed samples used to be the major deficiency of the test. In recent years the variable confined stress technique has been adopted, as a remedial measure.

b) The dynamic simple shear test has drawn much attention for its unique feature of simulating natural stress conditions of soil, and for applying relatively uniform horizontal shear stress to the soil sample throughout the whole process. This is an important factor for assessing soil liquefaction potential.

5. *Model test on the site and prototype observation*

A model test here denotes mainly the dynamic loading test on foundation and underlying soil which provides the dynamic rigidity, dynamic bearing capacity and dynamic modulus of deformation, etc.

Table 2.4.1. Dynamic testings and their functions.

Dynamic strain	10^{-6}	10^{-4}	10^{-2}	10^{0}
Frequency (hz)	10^{2}	10	10^{0}	10^{-1}
Mechanical characters	elastic, visco-elastic		elasto-plastic	stability dynamic effect & response
Parameters/or Dyn. Properties	Wave velocity, dynamic modulus, dynamic rigidity, Poisson's ratio, damping ratio, dynamic strength (c, ϕ), seismic liquefaction			
Lab tests: Dyn. triaxial, dyn. simple shear				
Resonance column				
Shaking table centrifuge				
In-situ tests: Wave velocity measuring				
dynamic model test				
micro-tremour measuring				
Prototype observation monitoring				
Actions to be simulated	Various vibration and wave actions	Dynamic deformation, seismic settlement and ruptures		Foundation failure, liquefaction, landslide, colluvial

Prototype observation here denotes mainly the monitoring of the behaviour of using instrumentation such as the seismograph. When a shock occurs, the actual response and behaviour of the structure can be obtained. Even without sensible shocks, some records reflecting microtremors can still be made.

6. *Measuring microtremor on the ground*
Microtremor measuring is suggested to be done on both the outcrop and the ground soil. The former can be used as the typical data representing local district background. The latter may be used to reflect the magnification and filtering characteristics of the soil layer in comparison with the bedrock.

A summary of the above-mentioned test methods is given in Table 2.4.1. The functions and applicable ranges of strains for each method are listed in this table.

2.4.3 *General requirements of all the assessment methods*

Different approaches may lead to different or even opposite results. If this is the case, more methods of assessment have to be applied. Among all the methods and criteria available, the most practical and advisable approach is the one strongly related to experience derived from historical events. Integration of theoretical analysis with practical experience of similarity, and realistic dynamic parameters of ground soils may produce reliable and satisfactory results.

CHAPTER 3

Earthquake ground motion

Earthquake ground motion can be simply called earthquake motion or ground motion. It is used to express the earthquake action at a site by means of concrete physical concepts, a series of parameters and their numerical models. Such action forms the basis of seismic design.

3.1 CHARACTERISTICS OF EARTHQUAKE GROUND MOTION

Figure 3.1.1 shows a typical time history of acceleration, velocity and displacement of earthquake ground motions. Observations based on many real records show that:

1. The acceleration component of ground motion is rich with high frequencies, and the displacement component is rich with long periods.

2. Some strong motions may last several tens of seconds, but some last only several seconds.

3. In some ground motion records, the maximum peak acceleration occurs once or twice and is accompanied by much smaller minor peak acceleration. However, in some other records, the maximum and minor accelerations are of comparable magnitudes and occur many times in one record.

4. Some earthquakes contain long major periods while others contain shorter periods.

For adequate seismic design, the above observations suggest that ground motion should be characterized by three major elements: ground motion intensity, frequency and time duration. These aspects are disccussed in details in the following.

3.1.1 *Ground motion intensity*

Ground motion intensity describes the magnitude of ground motion at a particular location and is expressed in terms of amplitude of acceleration, velocity or displacement.

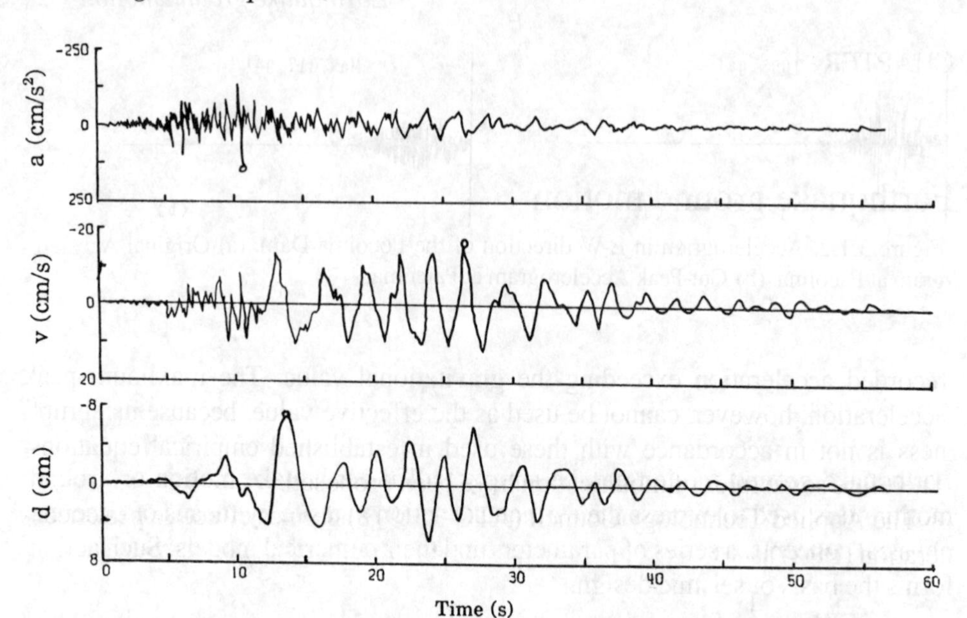

Figure 3.1.1. A typical and calibrated accelerogram and velocity, displacement time histories.

1. *Peak acceleration and its maximum value*
Figure 3.1.1 shows a typical calibrated accelerogram with the maximum peak values marked by a small circle. The velocity diagram is obtained by integration over the accelerogram while the displacement diagram is obtained by integration over the velocity diagram.

The maximum peak acceleration is closely related to the high frequency component of ground motion. Existing accelerograms may accurately record vibrations of periods greater than 0.06 s. The accuracy will drop seriously for vibrations of periods less than 0.04 s. Therefore the value of maximum acceleration may be subjected to errors. However, very high frequency components (periods less than 0.01 s) have little significance to buildings and structures, because of three reasons. The first reason is that such high frequency vibrations are filtered out by the soil and foundation; the second is the high attenuation during wave propagation from the source to the structure and the third is such high frequencies are too far away from the natural frequencies of structures to cause resonance.

In view of these reasons, the term 'effective peak value' is used to specify those accelerations which have practical influence on the dynamic response of buildings and structures.

2. *Effective peak acceleration*
Figure 3.1.2 shows the accelerogram in the E-W direction of the Pacoima dam obtained during the San Fernando earthquake in 1971. This is the first time of

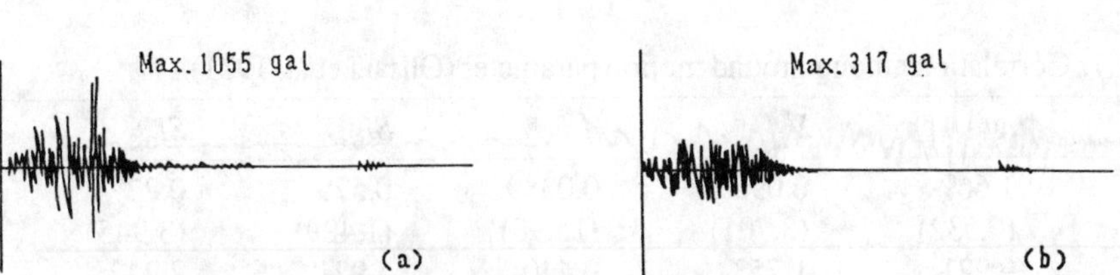

Figure 3.1.2. Accelerogram in E-W direction of the Pocoima Dam. (a) Original Accelerogram at Pacoima, (b) Cut-Peak Accelerogram at Paicoma.

recorded acceleration exceeding the gravitational value. The maximum peak acceleration, however, cannot be used as the effective value, because its abruptness is not in accordance with these used in established empirical equations. Therefore some rules of modification have been proposed.

The Applied Technology Council (ATC-3, 1978) defines effective peak acceleration (EPA) as

$$\mathrm{EPA} = \frac{1}{2.5} S_a$$

S_a is the average value of the acceleration response spectrum (damping ratio is 0.05) within the period of 0.1 ~ 0.5 s.

Thus, EPA is correlated with the real peak value, but not equal to nor even proportional to it. If the ground motion consists of high frequency components, EPA will be obviously smaller than the real peak value. The empirical constant 2.5, is essentially an amplification factor of the response spectrum obtained from real peak value records.

3. *Correlations among ground motion parameters*

Correlations among ground motion parameters provide an effective means to determine the design earthquake. Many researchers have performed regression analyses among the maximum values of acceleration (a), the velocity (V), the displacement (d) and the spectrum intensity (SI). An example given by Ohsaki et al. (1980) is shown in Table 3.1.1.

Two spectrum intensities have been defined by Housner (1952), one corresponding to a damping ratio of 0.0 ($SI_{0.0}$) and the other to 0.2 ($SI_{0.2}$). Notes for the values in Table 3.1.1 are:

1. Values along the diagonal are the ratios of X/Y.

2. Values above the diagonal are the ratio of the parameters for the horizontal direction. Values below the diagonal are the ratios of parameters for the vertical direction.

3. Values without brackets refer to ratios of a quantity in the horizontal direction to that in the vertical direction. Values within brackets refer to ratio of a quantity in the vertical direction to that in the horizontal direction.

Examples:

Table 3.1.1. Correlation among ground motion parameter (Ohsaki et al. 1981).

	a_H	V_H	d_H	$SI_{0.0}$	$SI_{0.2}$
a_V	1.669	0.0961	0.0389	0.579	0.227
	(0.532)	(8.701)	(15.98)	(1.429)	(3.945)
V_V	9.921	1.753	0.440	5.921	2.242
	(0.088)	(0.448)	(2.00)	(0.162)	(0.431)
d_V	17.67	1.839	1.555	11.82	4.290
	(0.046)	(0.476)	(0.468)	(0.071)	(0.181)
$SI_{0.0}$	1.551	0.157	0.0741	1.887	0.371
	(0.551)	(5.821)	(10.63)	(0.489)	(2.610)
$SI_{0.2}$	4.819	0.469	0.216	2.885	2.038
	(0.186)	(2.035)	(3.614)	(0.334)	(0.435)

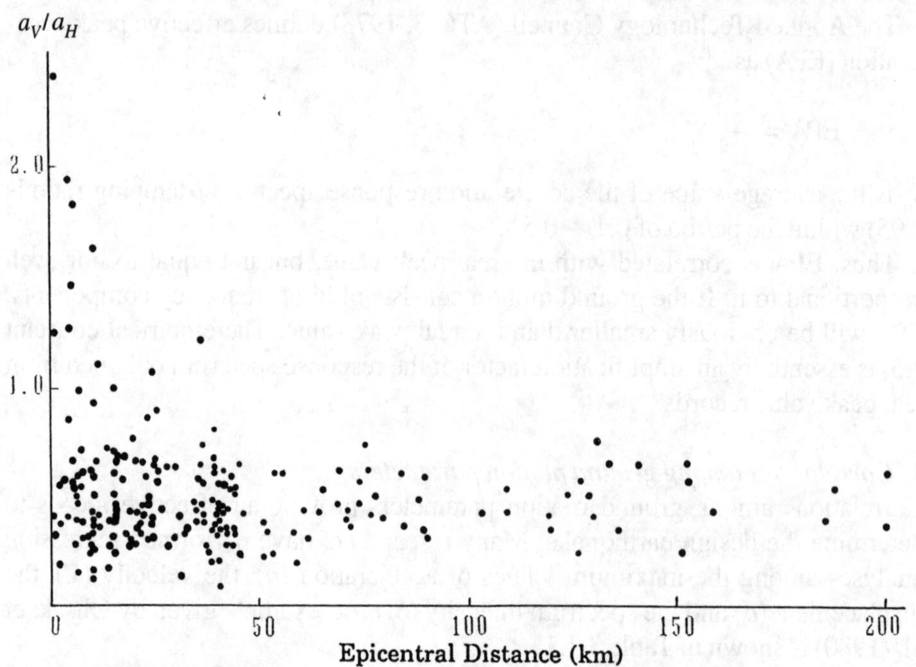

Figure 3.1.3. Relationship between average ratio (a_V/a_H) and epicentral distance.

$$a_H = 1.669\, a_V$$
$$a_V = 0.532\, a_H$$

In general, the average ratio a_V/ a_H is about ½ ~ ⅔ and varies with the epicentral distance as shown in Figure 3.1.3. The scatter of a_V/a_H values is larger for shorter epicentral distances. Hu (1988) points out: (1) a_V is always smaller than a_H for $a_H < 0.5$ g; (2) the average ratio of a_V/a_H varies between ½ ~ ⅔ with some scatter for the causative fault distance $D \leq$ 10 km; (3) a_V approaches a_H for $a_H \geq 0.5$ g.

Table 3.1.2. V_{max}/a_{max} values versus local lithology (Seed et al. 1976).

Epicentral distance (km)	Bedrock			Hard soil			Unconsolidated soil		
	a_{max}	V_{max}	$\frac{V_{max}}{a_{max}}$	a_{max}	V_{max}	$\frac{V_{max}}{a_{max}}$	a_{max}	V_{max}	$\frac{V_{max}}{a_{max}}$
	g	cm/s	in/s.g	g	cm/s	in/s.g	g	cm/s	in/s.g
15	0.42	34.0	28	0.38	54.0	56	0.27	54.0	80
20	0.33	22.0	22	0.30	34.0	45	0.22	34.0	62
40	0.105	8.0	25	0.115	11.5	40	0.105	11.5	43
60	0.05	4.4	29	0.063	6.0	38	0.065	6.0	37
Average			26			45			55

Seed et al. (1976) obtain values of V_{max}/a_{max} based on historical records of events with $M > 6.5$ in Western US and shown in Table 3.1.2 where V_{max} and a_{max} are the maximum velocity and acceleration, respectively. The values are dependent on lithology and epicentral distance.

3.1.2 *Frequency characteristics of ground motion*

Ground motion is composed of numerous components of different frequencies varing with site conditions. It is not a harmonic motion, but a stochastic vibration or random vibration of varying amplitudes and frequencies. However, for a certain earthquake, the ground motion can be treated as a motion composed of a series of simple harmonic waves of different frequencies. The diagram showing the amplitude variation of a parameter of the ground motion versus the corresponding period or frequency is called a spectrum. Three kinds of spectra are commonly used in earthquake engineering: Fourier amplitude spectrum, responses spectrum, and power spectrum. Among them, the response spectrum is the most common and may reflect the dynamic response of a structure to a seismic action.

1. *Spectral parameter and normalized spectrum*

Any response spectrum calculated from ground motion record can be normalized with its respective ground motion parameter. Within a specified range of period, the normalized response spectra with respect to the peak value of acceleration, velocity or displacement of ground motion, i.e. $S_a(T)/a$, $S_V(T)/V$, $S_d(T)/d$ have been considered.

In practice, the normalized spectrum with respect to peak acceleration is most commonly used. In China, the dynamic amplification factor (DAF) is proposed to express the amplified horizontal acceleration spectrum of a single mass system. It is defined as:

$$\mathrm{DAF} = \frac{S_a(T)}{a_H}$$

The DAF reflects the shape of the spectrum and not the amplitude. One simple expression of the shape is by means of the predominant period, namely the controlling period at which the maximum peak value occurs. As shown in Figure 3.1.4, the predominant period corresponds to the period of 0.5 seconds. Because of the multi-peak nature of this record other periods such as 0.35, 0.79, 1.1 seconds may also also be included as the predominant periods based on engineering judgement. These additional periods will provide the designer an important information to avoid possible phenomenon of resonance occurring in the designed structures.

2. *Limit of response spectrum*

The response spectrum is often expressed as a combination of the linear segments derived from maximum acceleration, maximum velocity and maximum displacement. The ratios V/a and aD/V^2 can be used to verify the limits of the spectrum with respect to two transfer periods (T_1 and T_2 as shown in Figure 3.1.5) which

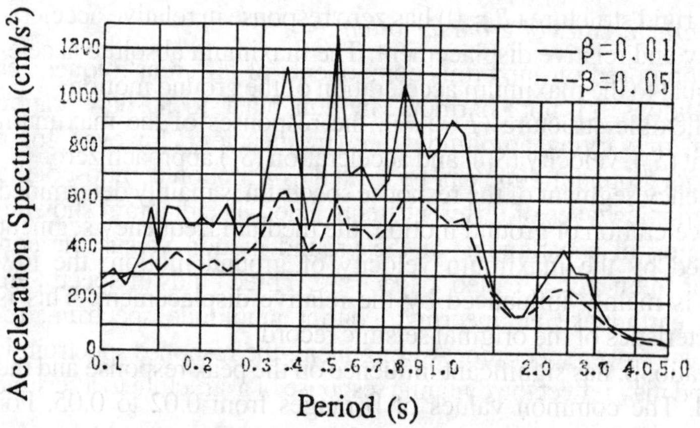

Figure 3.1.4. Spectrum recorded at Tokachi-oki, Japan in 1968.

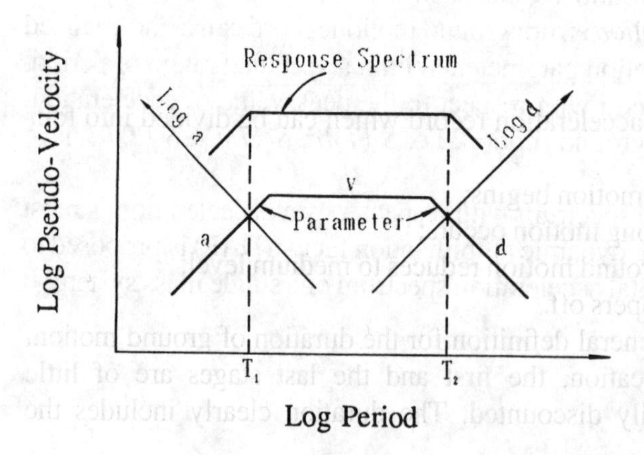

Figure 3.1.5. Sketch of transfer periods (T_1 and T_2) of spectral curve.

separate acceleration, velocity and displacement based on the following relationships (Newmark & Hau 1969):

$$T_1 \propto \frac{V}{a}$$

$$\frac{T_2}{T_1} \propto \frac{aD}{V^2} \tag{3.1.1}$$

3. *Characteristics of response spectrum*

Response spectrum theory is based on the following assumptions:

1. The seismic response of the structure (or medium) is elastic;
2. The ground motion at all the supports of the structure are equal and there is no interaction between the foundation and the ground;
3. The maximum seismic response is the most unfavourable response displayed by the structure.

The response spectrum is characterized by the following:

1. The absolutely rigid structure ($T = 0$) has zero response in relative acceleration, relative velocity and relative displacement. The maximum absolute acceleration response is equal to the maximum acceleration of the ground motion.
2. For infinitely flexible structure ($T \to \infty$), the responses of the maximum relative displacement (S_d), velocity (S_V), and acceleration (S_a) approach zero.
3. The high frequency segment of the response spectrum is mainly determined by the maximum acceleration of ground motion; the medium frequency segment is mainly determined by the maximum velocity of ground motion; the low frequency segment is mainly influenced by the relative displacement. This is similar to the characteristics of the original seismic record.
4. The damping ratio, β, has significant influence on the peak response and the predominant period. The common values of β ranges from 0.02 to 0.05. For $\beta = 0.02$, the peak response is reduced to one half of that for β approaching zero.

3.1.3 *Duration of ground motion*

1. *Basic concept*

Figure 3.1.6 shows a typical acceleration record which can be divided into four stages:

1. During 0 ~ 4 s: seismic motion begins;
2. During 4 ~ 12 s: the strong motion occurs;
3. During 12 ~ 38 s: the ground motion reduces to medium level;
4. After 38 s: the motion tapers off.

It is difficult to establish a general definition for the duration of ground motion. For most engineering application, the first and the last stages are of little significance and are normally discounted. The duration clearly includes the

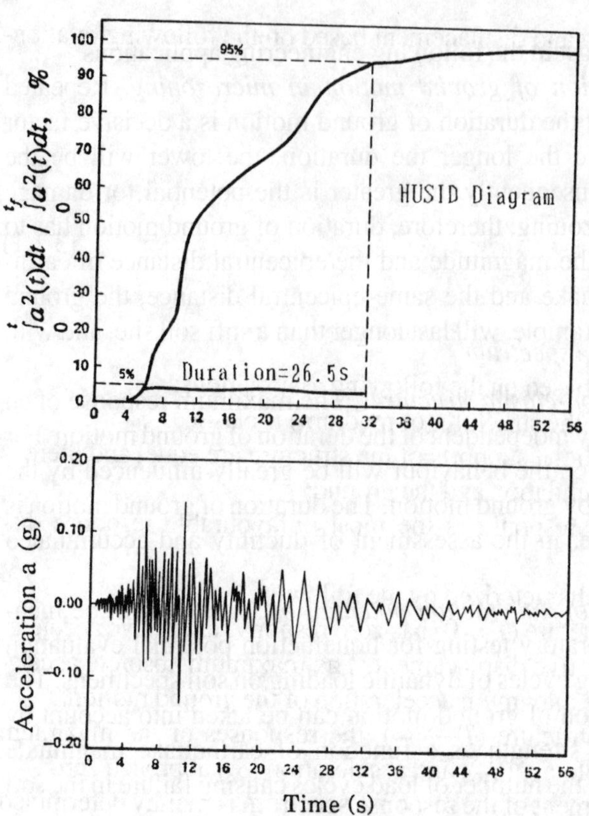

Figure 3.1.6. Typical HUSID diagram, duration definition, by Trifunac & Brady (1975), Dobry (1978).

second stage of strong motion and part of the third stage of medium motion. How large a part of the third stage should be included depends on the specific project under consideration.

Hence, different definitions for the duration of ground motion exist for different types of application. Two approaches of definition have commonly been used. The first approach defines the duration in terms of absolute amplitude of acceleration. It is the time elapsed between the first and the last amplitudes of acceleration equal to or greater than 0.05 g. This approach is based on the experience that such ground acceleration might be destructive.

The second approach defines the duration by relative amplitude of acceleration. It is the time elapsed between the first and the last acceleration amplitudes equal to or greater than $\frac{1}{2}a$, $\frac{1}{3}a$ or $\frac{1}{e}a$, where a is the measured maximum acceleration and e is the natural logarithmic base.

There are other ways to define the duration. For example, there is one defined by the relative energy or average energy of ground motion. However, these definitions are less commonly used than the above two definitions.

2. *Application of duration of ground motion to engineering practice*

The duration of ground motion is one of the most important parameters used for seismic design. It is used at least in the following engineering applications:

a) *Consideration of duration of ground motion in microzoning*. Repeated observations have shown that the duration of ground motion is a decisive factor for ground damages, because the longer the duration, the lower will be the ductility of structures and consequently the greater is the potential for damage during earthquakes. In microzoning, therefore, duration of ground motion has to be considered in addition to the magnitude and the epicentral distance of earthquakes. For the same earthquake and the same epicentral distance, the ground motion at a soft soil site, for example, will last longer than a stiff soil site, and will, therefore, cause more damage.

b) *The seismic design of non-elastic structures*. The maximum response of an elastic structure is theoretically independent of the duration of ground motion. For non-elastic structures, however, the behaviour will be greatly influenced by the duration of vibration induced by ground motion. The duration of ground motion is an important factor, therefore, in the assessment of ductility and accumulated damage of structures.

c) *Laboratory assessment of liquefaction potential of soils*. As will be mentioned in Chapter 7, the laboratory testing for liquefaction potential evaluation may be carried out by applying cycles of dynamic loading on soil specimens. The important effect of the duration of ground motion can be taken into account by considering the liquefaction strength as a function of earthquake magnitude which in turn can be related to the number of load cycles causing failure in the soil specimen.

d) *Influence of duration on earthquake intensity*. Earthquake intensity is mainly a measure of the degree of earthquake damages on man-made structures. It cannot be determined solely by the maximum peak horizontal acceleration. Some earthquakes may exhibit maximum accelerations as high as 0.2 ~ 0.7 g on the ground surface but only cause very limited damage due to the short durations of ground motion. On the contrary, other earthquakes with acceleration lower than 0.2 g have caused heavier damages because of long duration (say 30 ~60 s) of ground shaking.

3.2 CORRELATIONS BETWEEN EARTHQUAKE INTENSITY AND GROUND MOTION PARAMETERS

Earthquake intensity has three main applications:

1. As an indicator of overall earthquake damages.

2. To provide earthquake engineers a macroscopic measure of seismic influence of historical events. These events give important clues to earthquake activities, recurrence relationship and seismic design.

3. As a comprehensive index summarizing existing empirical experience for zoning purpose and for specifying relevant parameters for seismic design.

The need for seismic calculations leads to the establishment of numerous correlations between earthquake intensity and ground motion parameters. Among them, the relationships between intensity (I) and peak acceleration (a) are most frequently used and are summarized in Table 3.2.1.

In addition, correlations between intensity (I) and other parameters of ground motion such as horizontal velocity (V_H) and displacement (d_H) have also been investigated and summarized by Trifunac et al. (1977) as shown below:

$$\lg V_H = 0.25\,I - 0.65 \qquad \text{for } I = \text{IV} \sim \text{X}$$

$$\lg V_H = [0.29 \pm 0.01]I - [0.93 \pm 0.23]$$

$$\lg d_H = 0.191\,I - 0.53 \qquad \text{for } I = \text{V} \sim \text{X}$$

where I is in Modified Mercalli scale and a numeric value corresponding to the Roman numerials should be substituted into the equations.

In the above equations, V_H is in cm/s and d_H is in cm. The different equations are originally by different authors and choice on which one to use depends on the conditions specified for the equation and on the preference of the engineer.

McGuire et al. (1977) proposes relationships correlating a certain ground

Table 3.2.1. Correlations between earthquake intensity (I) and peak ground acceleration (a) [1].

Researcher	Intensity (*MM*)	Acceleration (cm/s^2)	Suggested formula
Hesshberger (1956)	III-VII	1-300	$\log a = \frac{3}{7} I_{MM} - \frac{9}{10}$
Ambraseys (1974)	IV-VII	2-600	$\log a_{\max} = 0.36\,I_{MM} - 0.16$
Trifunac, Brady (1975)	IV-X	7-1150	$\log a_k = 0.30\,I_{MM} + 0.014$
Murphy, O'Brien (1977)	I-X	10-1100	$\log a_k = 0.25\,I_{MM} + 0.25$
Institute of Eng. Mechanics, China (1973)	III-VIII	1-490	$\log a_{\max} = 0.35\,I_{MM} - 0.75$

Table 3.2.2. Physical parameter A (a, V, d) versus M, Δ and I ($\ln A = C_1 + C_2M + C_3 \ln \Delta + C_4 I$).

Physical parameter	C_1	C_2	C_3	C_4	$\sigma_{\ln A}$
a (cm/s^2)	2.71	–	–	0.601	0.781
	2.01	–	−0.313	0.506	0.723
	1.81	0.904	0.901	–	0.696
V (cm/s)	−1.51	–	–	0.543	0.770
	−1.11	–	−0.72	0.521	0.771
	−1.58	0.997	−0.710	–	0.715
d (cm)	−1.47	–	–	0.415	0.790
	−2.35	–	0.157	0.463	0.780
	−2.67	0.863	−0.398	–	0.740

Table 3.2.3. Peak ground acceleration (a) versus intensity (I) and magnitude (M) (Hu et al. 1983).

Magnitude (M)	Intensity (I) VI	VII	VIII	IX	X
5.5	0.135	–	–	–	–
6.0	0.082	0.284	–	–	–
6.5	0.048	0.179	0.583	–	–
7.0	0.030	0.104	0.389	1.154	–
7.5	0.022	0.064	0.228	0.833	–
8.0	0.018	0.044	0.136	0.502	1.747
8.5	0.017	0.034	0.091	0.292	1.100

motion parameter $A(a, V, d)$ with earthquake intensity (I), magnitude (M) and epicentral distance (Δ). Table 3.2.2 is a comprehensive result as an example.

Hu & Zhang (1983) correlate peak ground acceleration (a) with intensity (I) and magnitude (M) based on the data obtained from St. Andreas earthquakes along the west coast of the United States. The resulting relationships are shown in Table 3.2.3.

3.3 FACTORS INFLUENCING GROUND MOTION CHARACTERISTICS

Factors influencing ground motion can be classified into three categories:

1. *Seismic source*

The seismic source is characterized by the stress condition and mechanism of the earthquake, size of faulting, focal depth, pattern of dislocation, mode of radiation from the source, etc.

2. *The path of seismic wave propagation*

The path of seismic wave propagation is subject to geometric dispersion, absorption, heterogeneity of transmitting media (e.g. intrusive rock, fractured zone, etc.), transformation from body wave to surface wave, etc. In most engineering practice, however, the epicentral distance is considered as an approximation.

3. *Local site conditions*

Local soil conditions, topographic and geomophological features and the interaction between ground and superstructure may play a decisive role. The soil conditions may modify the distribution of seismic energy of different frequency components. The topographic and geomorphological factors may influence the distribution of seismic energy in spacial geological media.

3.3.1 *Factors influencing ground motion intensity*

As mentioned in Section 3.1.1, the ground motion intensity is expressed in terms of maximum acceleration, velocity or displacement. However, the maximum acceleration is the most commonly used parameter in earthquake engineering. Factors affecting the ground motion intensity or maximum acceleration can be demonstrated by the data of the Itajima bridge, Japan, during different tremors (Table 3.3.1). The data show:

Table 3.3.1. Acceleration records at Itajima Bridge, Japan (Ohsaki 1980).

Earthquake	Date	Magnitude (M)	Epicentral distance (km)	Max. ground accel. (gal)		Max. accel. on bridge pier (gal)	
				Longit.	Transv.	Longit.	Transv.
Hyuganada	1968.4.1	7.5	100	175	156	219	310
Hyuganada*	1968.4.1	6.3	100	35	42	39	66
Bungo-suido	1968.8.6	6.6	11	453	365	198	239

* After shock.

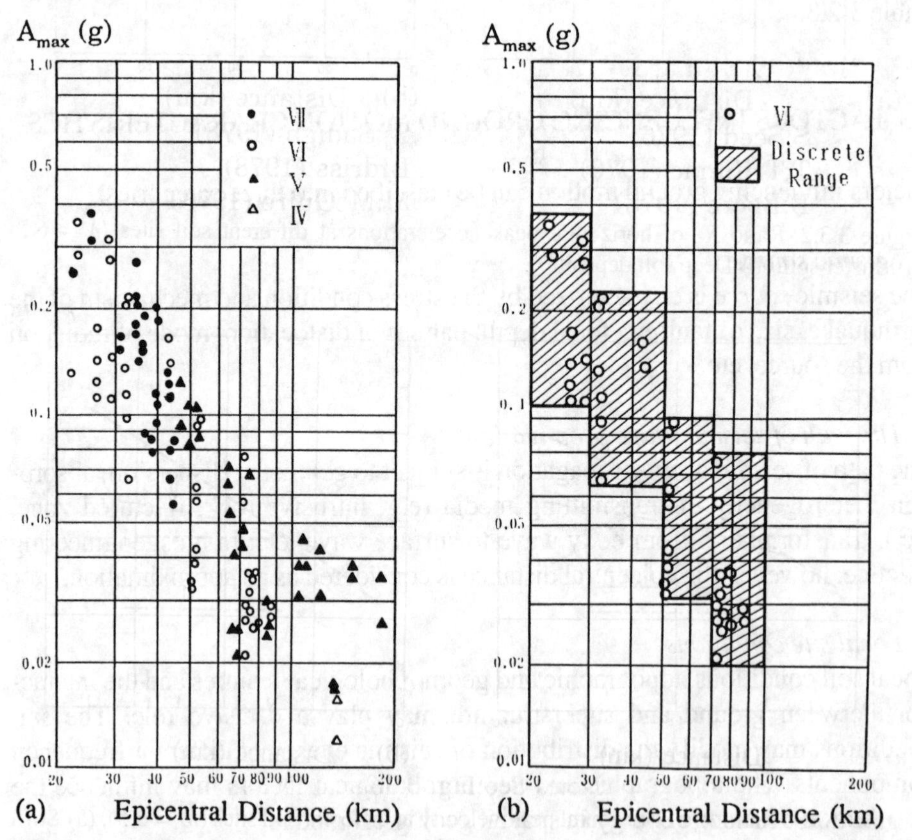

Figure 3.3.1. Data of A, I, Δ during San Fernando Earthquake. (a) Overall survey, (b) Specific to I = VI, for example.

1. For the same site and epicentral distance, the maximum ground acceleration increases with the magnitude (*M*).

2. For the same site and *M*, the maximum acceleration increases with decrease in epicentral distance.

Figure 3.3.1 show the real records of the acceleration versus epicentral distance in different earthquake intensity zones based on MMS during the San Fernando earthquake. The data clearly show that the maximum acceleration decreases with epicentral distance.

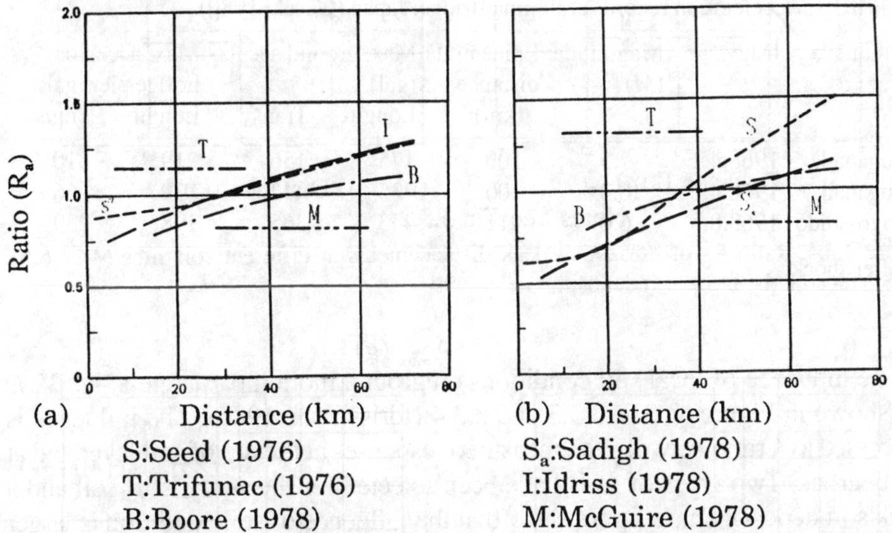

Figure 3.3.2. Ratio R_a of horizontal peak accelerations at different soil sites ($M = 6.5$). (a) Stiffer soil, (b) Deep soft deposit.

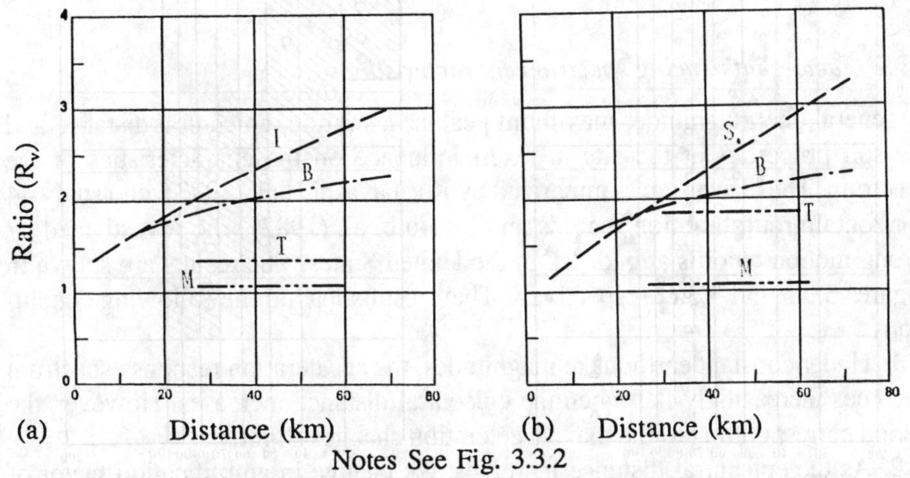

Notes See Fig. 3.3.2

Figure 3.3.3. Ratio R_v of horizontal peak velocity at different soil sites ($M = 6.5$). (a) Stiffer soil, (b) Deep soft deposit.

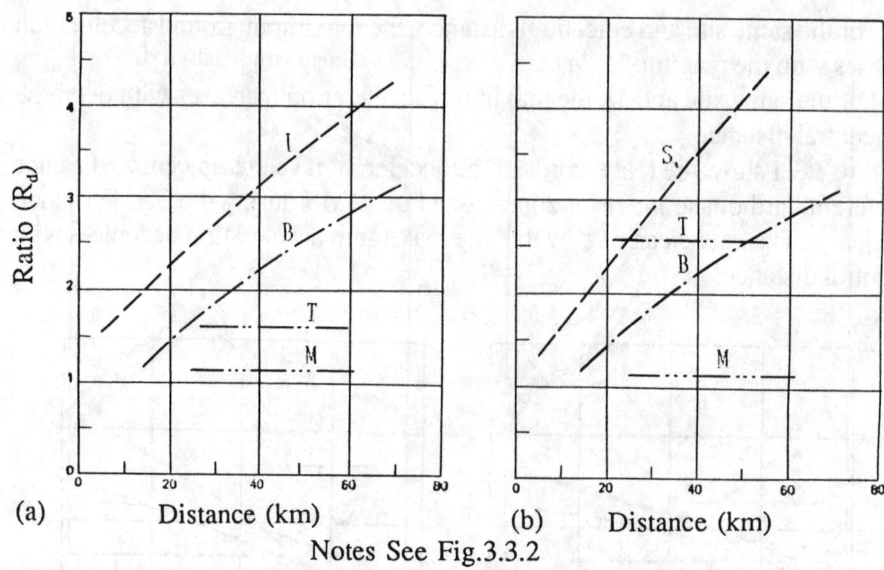

Figure 3.3.4. Ratio R_d of horizontal peak displacements at different soil sites (M = 6.5). (a) Stiffer soil, (b) Deep soft deposit.

The influence of local soil conditions on ground motion parameters – a, V, D are shown in Figures 3.3.2, 3.3.3 and 3.3.4 (Idriss et al. 1985). The influence is expressed as a ratio between the measured responses at the top of soil layer and at the bedrock. Two types of soils have been selected for the study: stiff soil and a thick soft deposit. These figure show that the influence of soil condition is larger in displacement, but smaller in acceleration. The peak acceleration at bedrock in the near field is greater than that at soft soil. However, in the far field, the peak acceleration at the soft soil is greater than that at the bedrock.

3.3.2 *Factors influencing spectrum characteristics*

In general, the magnitude, maximum peak acceleration, epicentral distance and the soil properties at the site all exert influence on the characteristics of the spectrum. The effects are summarized by Kuribayashi et al. (1973) who study 44 horizontal ground motion records and by Hu et al. (1982) who investigated 27 strong motion records at bedrock in the United States. The results are shown in Figures 3.3.5 and 3.3.6, respectively. Their results lead to the following conclusions.

1. Under constant earthquake magnitudes, the acceleration response spectrum becomes increasingly flat when the epicentral distance increases. However, the period corresponding to the max. acceleration changes slightly.

2. As the epicentral distance increases, the change in amplification factor of acceleration is small for short periods but change is large for long periods.

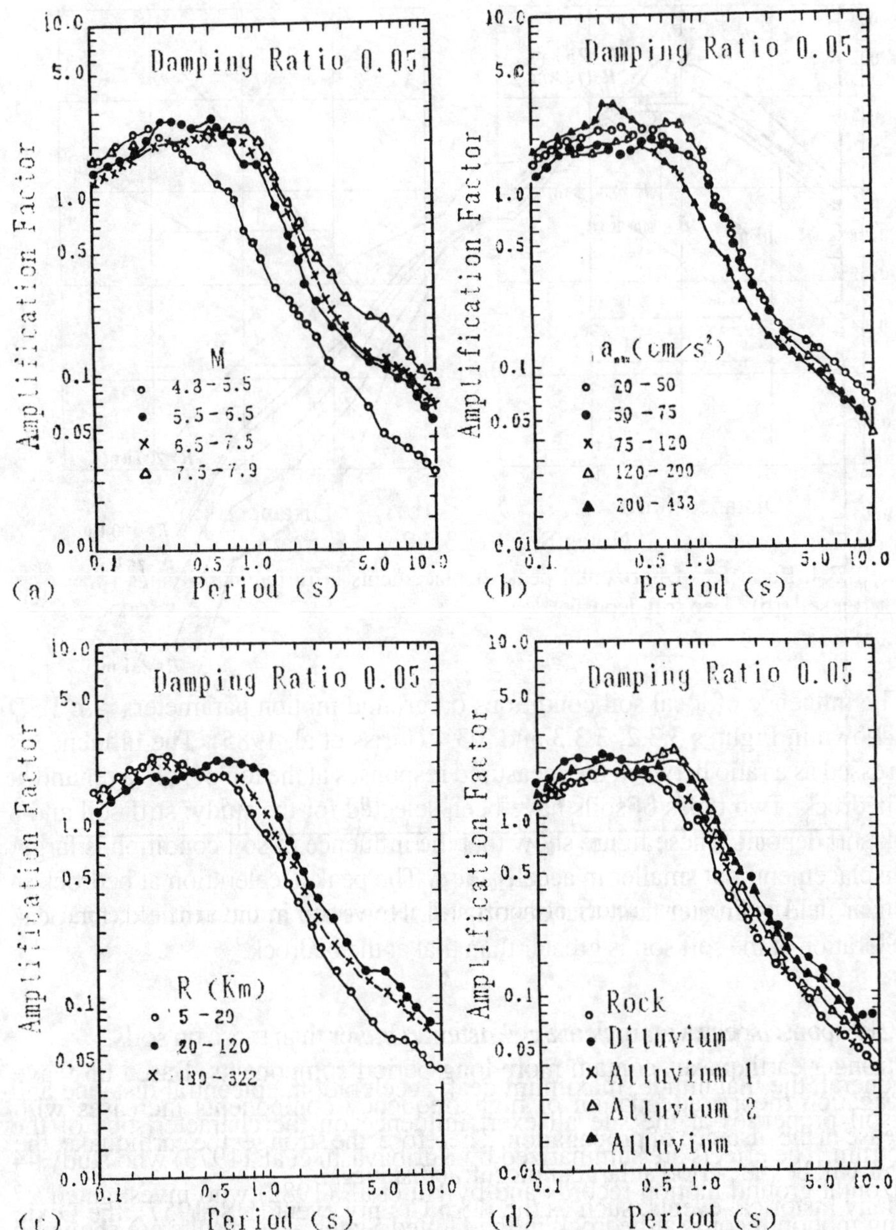

Figure 3.3.5. Influence of (a) Magnitude, (b) Peak accelerations, (c) Epicentral distance and (d) Soil condition on spectra.

3. For the same epicentral distance, the acceleration response spectrum increases with increasing magnitudes. The increase is again larger at long periods than at short periods.

4. The period corresponding to the maximum acceleration changes slightly when the magnitude varies.

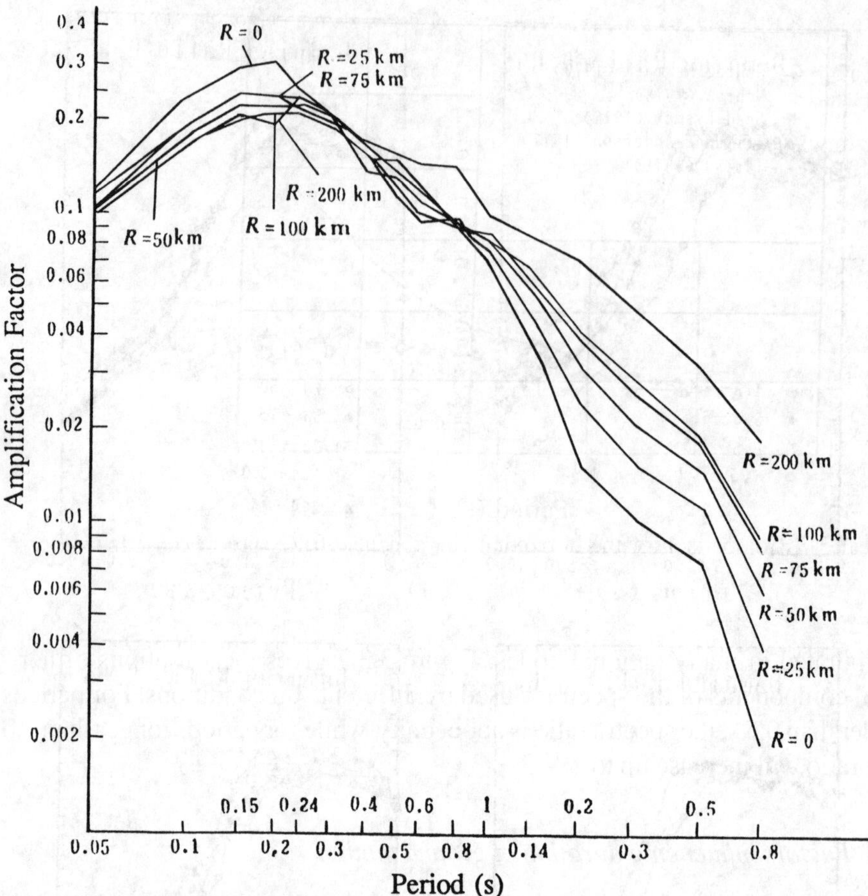

Figure 3.3.6. Amplification factors of horizontal acceleration at different epicentral distances.

5. Response spectra on rock are consistently lower than those on soils.

Stronger earthquakes contain more long period components. Based on wave propagation theory, attenuation of high frequency components increases with increase in the distance of propagation. Therefore, the stronger the earthquake, the longer will be the period at larger epicentral distances.

Many historical events, such as the Kern County event (US, 1957), the Dixie Valley and Fairview Valley earthquake (US, 1954), the Geiz earthquake (Turkey, 1970), the Guerrers earthquake (Mexico, 1957) and the Tangshan earthquake (China, 1976), etc. do reveal the fact that stronger earthquakes destroy many long period structures in the far field, especially those situating on soft ground, but cause less damage to short period structures in the same area. On the other hand, damages are concentrated in short period structures in the near field for earthquakes of medium magnitudes.

The influence of soil conditions on the spectrum has been recognized and

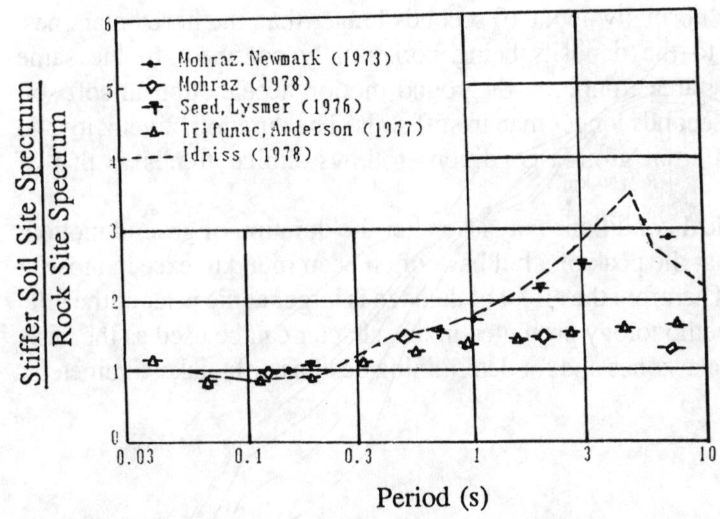

Figure 3.3.7. Comparison of the horizontal components of the spectra caused at different sites.

incorporated in many national codes. Figure 3.3.7 gives the variations of horizontal components of the spectra caused by different soil conditions. For periods smaller than 0.5 s, the spectra ratio is about unity, while for periods longer than 1.0 s, the ratio will increase up to 2 ~ 3.

3.3.3 *Factors influencing duration of ground motion*

The duration of ground motion is composed of three elements:

1. Duration related to source which can be approximately represented by the magnitude;
2. Duration related to propagation path or the epicentral distance;
3. Duration related to after effects of the vibration process which can be considered under the topic of local soil conditions.

However, most existing equations for calculating duration consider only the magnitude (or intensity) and the epicentral distance. Neglecting the effects of local soil conditions may lead to serious problems in properly understanding the factors affecting the duration. It is emphasized here that local soil conditions should be addressed.

Duration related to local soil conditions is mainly controlled by multi-reflection of seismic waves up and down between the bedrock and the ground surface. This effect is extremely important in controlling vibrations in loose and soft deposits transmitted from the bedrock. By inputting white noise to silt deposits of different thickness and shear wave velocities, Hakuno et al. (1975) show that the softer the soil, and the lower the velocity, the longer is the duration. Trifunac & Brady (1975) and Trifunac & Westermo (1977) also show that the durations of vertical ground motions are normally about 10 seconds longer than the horizontal ones. This may be due to the deposits being horizontally confined. In the same earthquake intensity areas, duration of ground motion acceleration in soft soil used to be 10 ~ 12 seconds longer than in stiff soils. The duration of velocity and displacement in different ground conditions follows similar trends as that of acceleration.

Local topographical conditions may also alter the duration of ground motion. Durations recorded at the peak of a hill have often been found to exceed those at the foot of the hill. Therefore the effect of damage is larger at the peak of the hill.

The topics and methodology presented in this chapter can be used as the basis for siting in earthquake zones and for determining design earthquake parameters.

CHAPTER 4

Seismic hazard analysis for a site

Seismic hazard analysis is to predict the influence of a future earthquake of certain magnitude on a site of interest. The risk originates from two aspects: the surrounding seismic zone of the potential seismic source and the site itself. The assessment of hazard analysis normally involves three main steps:

1. Locate the potential seismic source or sources surrounding the site and estimate its activity.
2. Ascertain the path of seismic wave propagation and its attenuation characteristics.
3. Adopt an appropriate model for seismic hazard analysis.

4.1 METHOD OF SEISMIC HAZARD ANALYSIS

4.1.1 *Deterministic method*

In principle, the deterministic method describes an independent variable in terms of a single numeral or a correlation of variables by establishing equations composed of certain parameters. In practice, the deterministic method is used to determine the intensity, peak acceleration of ground motion and the like on the site, based on either theoretical or empirical quantitative relationship such as the intensity-magnitude-epicentral distance equation, the magnitude-acceleration-displacement relationship, etc. The method produces a deterministic assessment of the seismic hazard at the site.

The conventional steps adopted for deterministic analysis are as follows:

1. Locate all the potential seismic sources geographically related to the site, based on the study of historical events and available geotectonic information.
2. Select the controlling earthquake characterized with magnitude and epicentral distance for assessment. Normally one or two sets of most unfavourable events are to be adopted according to the frequency range of the major structures on the site.
3. Based on the selected magnitude (M) and epicentral distance (Δ), evaluate

the maximum ground motion parameters taking into account the appropriate attenuation relationship. In most cases, the horizontal acceleration is to be used as the design parameter for ground motion.

4. Apply appropriate modifications to the above mentioned parameters in order to accommodate local site conditions.

4.1.2 *Probabilistic method*

The probabilistic approach is to identify the seismic hazard of a site in terms of the magnitude or intensity having a probability of exceedance within a certain time period. This method considers various probabilities of ground motion under specified earthquake intensities or magnitudes regardless of whether or not the epicentre lies within the site. Therefore, it is a general and systematic evaluation of seismic hazard of a site.

The main contents of this approach are:

1. Based on the actual need of seismic consideration in design, estimate the probability of fundamental intensity or design earthquake intensity.

2. Determine the ground motion parameters of the site, especially the peak ground acceleration.

3. Specify an acceleration response spectrum with respect to a certain range of vibration period pertinent to the conditions of the site of interest. This spectrum is to be used as the earthquake input from the bedrock.

4. Determine the duration of ground motion and the time history of earthquake input.

The major steps of seismic hazard analysis are:

1. Zoning of potential seismic sources.
2. Determining parameters of seismicity.
3. Selecting models of occurrence of seismicity.
4. Evaluating the attenuation rule on ground motion including intensity.
5. Calculating the probability of exceedance with the site as the centre of an area of radius equal to of 200 ~ 300 km.

4.2 ANALYSIS OF SEISMICITY

Seismicity analysis refers to the assessment of time, space and intensiveness of future earthquake events specific to a district. Seismic hazard analysis is based on the following information:

1. Possible locations of future seismic epicentres.

2. Possible magnitudes and their recurring frequencies from the possible seismic sources.

3. Distances and directions between the site and the possible seismic sources.

4.2.1 *Necessary information for seismicity analysis*

1. *Historical earthquake information*
Historical records dating as far back as possible form the main basis for evaluating seismicity. The records should include:
- Time of the earthquake;
- Epicentral intensity or the maximum intensity;
- Local site intensity;
- Contours of equal intensity;
- Magnitude;
- Location of epicentre and focal depth.

2. *Instrument-based observational data*
Data obtained through instruments normally cover two parts: earthquake source information obtained from seismographs and ground motion parameters detected by strong motion recorders. Data normally obtained are:
- Geometrical epicentre and focal depth;
- Time of earthquake;
- Magnitude of major shock;
- Magnitude and distribution of aftershocks;
- Ground movement records of the site.

3. *Seismogeological information*
Seismogeological information provides the most important record about tectonic movements and traces of past strong earthquakes. Such information includes the tectonic fracture pattern, the size, depth and age of the fault and its history of activities and relationship to earthquake occurrence.

4.2.2 *Potential earthquake zone*

Potential earthquake zone generally denotes the area with the site as centre and a radius of 200 ~ 300 km where destructive earthquake may occur. The exact delineation of the extent of potential earthquake zone is faced with uncertainty, and the degree of which depends on available information. In general, the more the information, the less is the uncertainty and hence the smaller one can define the potential earthquake zones. The main methods of zoning are given below:

1. Locate the site on the map and circle an area with the site as centre. The radius of the circle depends on the importance of the project and ranges between 200 ~ 300 km. The seismicity in this area is to be investigated.

2. Evaluate the character of tectonic movement within this area and further delineate seismic zones of different intensity or magnitude.

3. Study the relationship between tectonic fracture, fault and seismicity in the area.

4. Conduct a similar analysis of the site. Any active tectonic fracture zone

which may likely influence the hazard analysis of the site should be singled out as an independent potential earthquake zone.

5. Utilize existing maps of seismic zoning based on basic intensity and hazard analysis.

4.2.3 *Determination of upper limit magnitude of the potential earthquake zone*

The upper limit of earthquake magnitude is the highest probable level of seismicity in the potential earthquake zone in terms of the energy release of a single event. Several methods have been used to define this upper limit.

1. *Determination based on historical records*
For areas with abundant historical information, the maximum magnitude of historical events may be taken for the upper limit. To allow for possible errors in the records, it is sometimes advisable to add 0.5 to the maximum recorded magnitude as the upper limit.

2. *Magnitude-frequency analysis*
For some districts, sufficient information may exist for establishing an earthquake recurrence relationship. Such a relationship provides rough estimates of the recurrence period for specific earthquake magnitude.

This approach, however, is not applicable to those areas with only a few historical events over a long period of time.

3. *Tectonic analogy*
The magnitude of earthquake is closely related to the size of causative fault in the tectonic fracture zone. The upper limit of the magnitude of potential earthquake zone can be inferred from a district of similar tectonic background. By comparing the geotectonic features, deep fault fracturing behaviour and structural characteristics, some estimate of the upper limit can be made.

4. *Rupture length of causative fault/fracture zone*
Some regression analyses have been conducted between earthquake magnitude and the rupture length of causative fault/fracture zone. More details are discussed in Chapter 6.

4.2.4 *Determination of seismicity parameters*

1. *Magnitude versus accumulated frequency of occurrence*
The magnitude, M, can be related to the frequency of occurrence by

$$\ln N = \alpha - \beta M \qquad (4.2.1)$$

where N is the recurrence number of earthquakes with a magnitude equal to or greater than M, α and β are coefficients of regression obtained by the least square method. α represents the seismicity level of the area of interest, while β denotes the ratio of the number of minor earthquakes to the number of major earthquakes. The mathematical expressions for α and β are as follows:

$$\alpha = \frac{1}{n}\left(\sum_{i=j}^{n} \ln N(M_i) + \sum_{i=j}^{n} M_i\right) \tag{4.2.2}$$

$$\beta = \frac{\sum_{i=j}^{n} \ln N(M_i) - n\sum_{i=j}^{n} M_i \ln (M_i)}{\sum_{i=j}^{n} M_i^2 - \left(\sum_{i=j}^{n} M_i\right)^2} \tag{4.2.3}$$

The cumulative probability distribution function of magnitude, $F(M)$, and the probability density function of magnitude, $f(M)$, can be calculated from:

$$F(M) = \frac{1 - \exp[-\beta(M - M_0)]}{1 - \exp[-\beta(M_u - M_0)]}, \quad \text{for } M_0 \le M \le M_u \tag{4.2.4a}$$

$$f(M) = \frac{\beta \exp[-(M - M_0)]}{1 - \exp[-\beta(M_u - M_0)]} \tag{4.2.4b}$$

where M_u is the upper limit of magnitude, M_0 is the lower limit of magnitude. Practically, M_0 is taken as 4 ~ 5, since smaller earthquakes will cause no damage.

Shoh & Morga (1979) proposed a bilinear relationship

$$\begin{aligned} \ln N(M) &= \alpha_1 - \beta_1 M, \quad \text{for } M_0 \le M \le M_a \\ \ln N(M) &= \alpha_2 - \beta_2 M, \quad \text{for } M_a \le M \le M_b \end{aligned} \tag{4.2.5}$$

where M_a is the magnitude corresponding to the intersection of the two lines, M_b is the cut-off upper limit magnitude in the bi-linear relationship.

2. *Average annual rate of earthquake occurrence*

The average annual rate of earthquake occurrence, R_{a0}, is defined as the number of earthquakes with magnitudes > M_0 occurring in a specific district during a period of one year. It can be calculated from

$$R_{a0} = \frac{N(M_0) - N(M_u)}{T} \tag{4.2.6}$$

where T is the length of time in years, $N(M_0)$ and $N(M_u)$ are obtained from the magnitude versus frequency relationship.

The average annual rate of earthquake occurrence is an important parameter for earthquake hazard analysis. Figure 4.2.1 shows earthquake records over 500 years

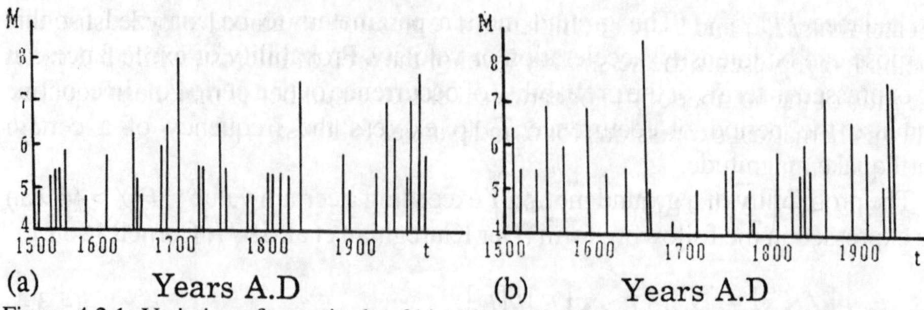

Figure 4.2.1. Variation of magnitude of historical seismic events, (a) in North China Plains zone, (b) in Tan-Lu fault zone [11].

for the North China plains and the Tan-Lu fault zone. R_{a0} for these two areas can be derived from such records using Equation (4.2.6).

4.3 SEISMIC HAZARD ANALYSIS

4.3.1 *Random process model of earthquake occurrence*

Poisson's model is the most popular and simplest process used for earthquake hazard analysis. This model is based on the assumption that two successive earthquakes are independent of each other with respect to the time and space of occurrence. The mathematical expected value of occurrence within a given time interval is constant and can be obtained from the magnitude-frequency relationship.

It is assumed in Poisson's model that the time interval between two successive events is independent and obeys the same distribution law. The probability density function and the distribution function are respectively:

$$f(t) = \lambda e^{-\lambda t} \tag{4.3.1}$$

$$F(t) = 1 - e^{-\lambda t} \tag{4.3.2}$$

The hazard function of Poisson's model, $\gamma(t)$ is given by

$$\gamma(t) = \frac{f(t)}{1 - F(t)} = \lambda \tag{4.3.3}$$

This is a constant independent of time and equal to the annual rate of occurrence.

4.3.2 *Numerical model of seismic hazard*

Seismic hazard analysis requires a given ground motion having a probability exceeding certain percentage at a site. It is also simply called the probability of exceedance $P[Y > y]$. The ground motion parameters to be analyzed for this purpose can be intensity, acceleration or velocity. Probability of exceedance can be expressed in terms of the probability of occurrence over a period of time, or one can use the period of recurrence T_r to express the frequency of a certain earthquake magnitude.

The probability of a ground motion Y exceeding a certain value y, $P[Y > y]$, can be expressed in the following form (Der Kiureghian et al. 1977; Cornell 1968):

$$P[Y > y] = \sum_{j=i}^{n} P[Y > y|E_j]P[E_j] \quad (4.3.4)$$

$$P[Y > y|E_j] = \int ... \int P_j[Y > y|x_1, x_2, x_3 ...]$$
$$\times f(x_1)\, f(x_2|x_1)\, f(x_3|x_2, x_1) ... \mathrm{d}x_3\, \mathrm{d}x_2\, \mathrm{d}x_1 \quad (4.3.5)$$

where Y = the ground motion parameter in consideration; y = the given value of the same parameter; j = the potential earthquake zone number; E_j = the seismic events having occurred in zone j; x_t (t = 1, 2, ...) = factors to be considered, such as magnitude, epicentral distance, length of fault, etc.; $f(x)$ = the probability density function; $P[E_j]$ = the probability of occurrence of seismic event E_j.

Seismic hazard analysis proceeds in two steps:

1. Determine the probability exceedance of a single event $P[Y > y|\ E_j]$.
2. Further determine the total probability exceedance of multiple events $P[Y > y]$, taking the possibility of repeated earthquakes into account.

Assuming there are n potential sources around the site, and let ν_j be the average annual rate of occurrence of events, with $M \geq M_0$, then the average annual rate of total occurrence is:

$$\nu = \sum_{j=1}^{n} \nu_j \quad (4.3.6)$$

The probability of occurrence of an event with $M \geq M_0$ in source zone j is:

$$P[E_j] = \frac{\nu_j}{\nu} \quad (4.3.7)$$

Therefore the probability of ground motion Y_j greater than a given value y induced by all the potential source around the site will be:

$$P[Y > y] = \frac{1}{\nu} \sum_{j=1}^{n} P[Y > y\ |E_j]\nu_j \quad (4.3.8)$$

Assuming occurrences of the seismic events in all the potential seismic source zones obey a homogeneous Poisson's process of ν, the annual probability of ground motion $Y > y$ is given by:

$$P_{1\,\mathrm{year}}[Y > y] = 1 - \exp\{-\sum_{j=1}^{n} \nu_j P[Y > y|E_j]\} \quad (4.3.9)$$

or for low probability of occurrence,

$$P_{1\,\text{year}}[Y > y] \simeq \sum_{j=1}^{n} \nu_j P[Y > y \mid E_j] \tag{4.3.10}$$

The corresponding period of recurrence T_r is

$$T_r = \left(\sum_{j=1}^{n} \nu P[Y > y \mid E_j] \right)^{-1} \tag{4.3.11}$$

The probability of exceedance in T years can be obtained by

$$P_T[Y > y] = 1 - \{1 - P_{1\,\text{year}}[Y > y]\}^T \tag{4.3.12}$$

In general, the calculation of probability of exceedance is based on the conditional probability $P[Y > y \mid E_j]$, namely the probability of ground motion $Y > y$ induced by a seismic event occurring in source zone j.

The probability of ground motion $Y > y$ induced by one seismic event in source zone j is

$$P[Y > y \mid E_j] = \int_{M_0}^{M_u} P_j[Y > y \mid E_{j,M}] f_j(M) \mathrm{d}M \tag{4.3.13}$$

where $E_{j,M}$ is a seismic event of magnitude M occurring at a distance R from the potential source in source zone j; $f_j(M)$ is the magnitude density function in the same zone.

The conditional probability $P_j[Y > y \mid E_{j,M}]$ is to be considered only for some causative fault rupture models such as by Liao et al. (1985).

There are three idealized patterns of potential seismic sources:

Source I: linear source with known orientation and location of the causative fault.

Source II: area source with known orientation and unknown location of the causative fault.

Source III: area source with both orientation and location of the causative fault unknown.

Source I

Figure 4.3.1 shows the case of a site situated on one side of a causative fault represented by a linear source with the symbols defined in the following: h is the focal depth; d is the horizontal projection of the distance between the site and the fault; s is the causative fault length corresponding to one earthquake event; L is the total length of existing fault and fractures; r_0 is the minimum distance between the site and the fault; r_y is the minimum distance between the site and the nearest end of the fault.

If the minimum distance r_y is smaller than the critical distance R obtained from the attenuation equation

$$R = g_1(y, M) \tag{4.3.14}$$

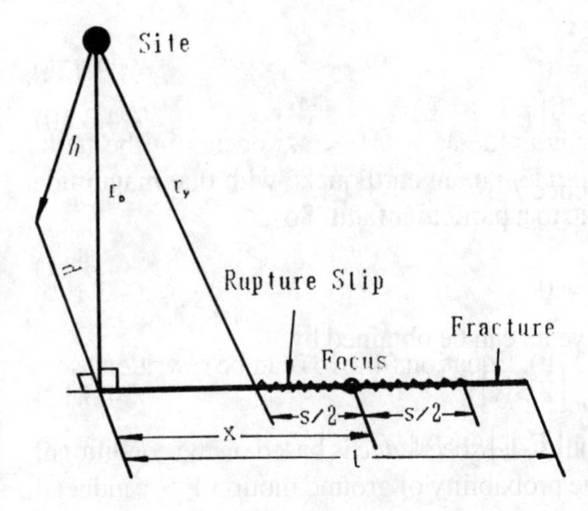

Figure 4.3.1. Source I – Particular case [12].

for the given magnitude M and motion y, the site may experience a motion larger than the given value y. On the other hand, if motion Y to be experienced at the site is smaller than the given value y, the effects of the given motion y on the site is small. The critical horizontal distance X corresponding to R is

$$X = \sqrt{R^2 - r_0^2} + \frac{s}{2} \tag{4.3.15}$$

where R is determined by Equation (4.3.14). By referring to Figure 4.3.1, one obtains:

$$P\left[Y > y \mid E_{j,M}\right] = \begin{cases} 0, & \text{if } R < r_0; \\ \dfrac{X}{L}, & \text{if } R \geq r_0, \quad \text{but } X < L; \\ 1, & \text{if } R \geq r_0, \quad \text{and } X \geq L \end{cases} \tag{4.3.16}$$

The probability of ground motion Y at the site exceeding the given value y can be obtained from Equation (4.3.13).

Let M_1' be the magnitude when the critical distance $R = r_0$ and be given by the attenuation equation

$$M_1' = g_2(a, r_0) \tag{4.3.17}$$

and M_0 is the lower limit of magnitude as mentioned before. From Equation (4.3.16), if $M_0 < M_1'$, the probability for a certain magnitude M to occur in source zone j is

$$P\left[Y > y \mid E_{j,M}\right] = 0$$

From Equation (4.3.13),

$$\int_{M_0}^{M'_1} P[Y > y \mid E_{j,M}] f_j(M) \mathrm{d}M = 0 \tag{4.3.17a}$$

if $M'_1 < M_0$, an earthquake with magnitude $M'_1 < M < M_0$ occurs in the fault. Since M_0 is also background magnitude, i.e. an earthquake with that magnitude can occur anywhere, it is not specific to a particular fault. So

$$\int_{M'_1}^{M_0} P[Y > y \mid E_{j,M}] f_j(M) \mathrm{d}M = 0 \tag{4.3.17b}$$

From Equations (4.3.17a) and (4.3.17b), Equation (4.3.13) can be rewritten as

$$P[Y > y \mid E_j] = \int_{M_1}^{M_u} P_j[Y > y \mid E_{j,M}] f_j(M) \mathrm{d}M \tag{4.3.18}$$

in which M_1 is obtained from:

$$M_1 = \max\{M_0, M'_1\}$$

In order to solve Equation 4.3.18, three cases have to be considered:

1. In the case that one end of the total fault is vertically below the site, there are four possibilities.

a) $M_1 > M_u$, this is an imaginary and impossible event, so

$$P[Y > y \mid E_j] = 0 \tag{4.3.19}$$

b) $M_1 < M_u$, but the given value y is so large that it can only happen when the length of the earthquake fault l equals the total fault length, i.e. $X \le L$ and $R \le r_0$. From Equation (4.3.16), the conditional probability is

$$P[Y > y \mid E_j] = \int_{M_1}^{M_u} f_j(M) \mathrm{d}M \tag{4.3.20}$$

since $P[Y > y \mid E_{j,M}] = 1$.

c) $M_1 < M_u$, and the length of earthquake fault is only one section of the total fault, i.e. $R \le r_0$ and $X < L$. From Equation (4.3.16), the conditional probability is

$$P[Y > y \mid E_j] = \int_{M_1}^{M_u} \frac{X}{L} f(M) \mathrm{d}M \tag{4.3.21}$$

d) $M_1 < M_u$, and the length of earthquake fault is only one section of the total fault, but M_u is greater than M_2 obtained from the relation of magnitude versus the length of earthquake fault, the conditional probability is

$$P[Y > y \mid E_j] = \int_{M_1}^{M_2} \frac{X}{L} f(M) \mathrm{d}M + 1 - F(M_2) \tag{4.3.22}$$

and M_2 is determined by

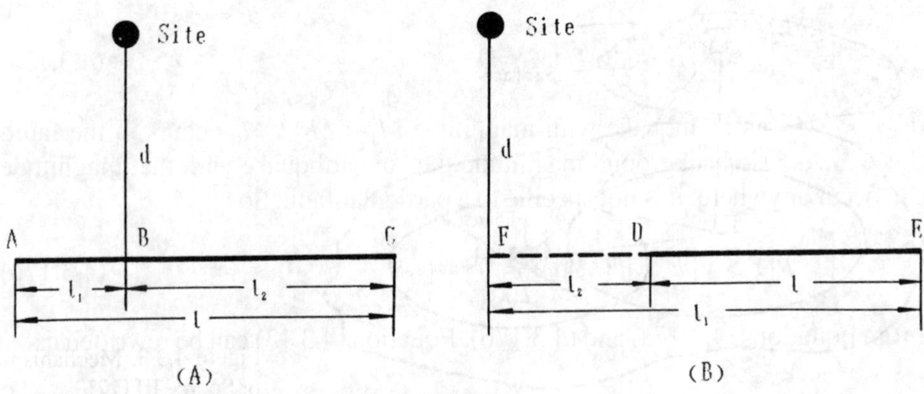

Figure 4.3.2. Source I (horizontal diagram) [12].

$$\sqrt{g_2^2(y, M_2) - r_0^2} + \frac{s}{2} = L \tag{4.3.23}$$

$M = g_1(y, R)$, $R = g_2(y, M)$ and $y = g(M, R)$ are the different forms of the same attenuation relationship.

2. In the case as shown in Figure 4.3.2A, the probability of $Y > y$ in source zone j is given by

$$P\left[Y > y \middle| E_j\right] = \frac{L_1}{L} P_{AB}\left[Y > y \middle| E_j\right] + \frac{L_2}{L} P_{BC}\left[Y > y \middle| E_j\right] \tag{4.3.24}$$

in which $P_{AB}\left[Y > y \middle| E_j\right]$ and $P_{BC}\left[Y > y \middle| E_j\right]$ are the probabilities relative to sections AB and BC, respectively, and are determined by the methods described above.

3. In the case as shown in Figure 4.3.2B, the probability of $Y > y$ in source zone j is given by

$$P\left[Y > y \middle| E_j\right] = \frac{L_1}{L} P_{DE}\left[Y > y \middle| E_j\right] - \frac{L_2}{L} P_{FD}\left[Y > \middle| E_j\right] \tag{4.3.25}$$

Source II

Source II can be divided into a number of strip zones for which treatment identical to that for Source I applies. The final answer is obtained by summing up all the strips to yield $P\left[Y > y \middle| E_j\right]$.

Source III

Figure 4.3.3 shows the model of Source III. The value of $P[Y > y | E_{j,M}]$ depends on the relative location of circle I with radius $\sqrt{r_y^2 - h^2}$ and circle II with radius $s/2$.

For different earthquake magnitudes, the value of $P[Y > y | E_j]$ can be obtained in the following:

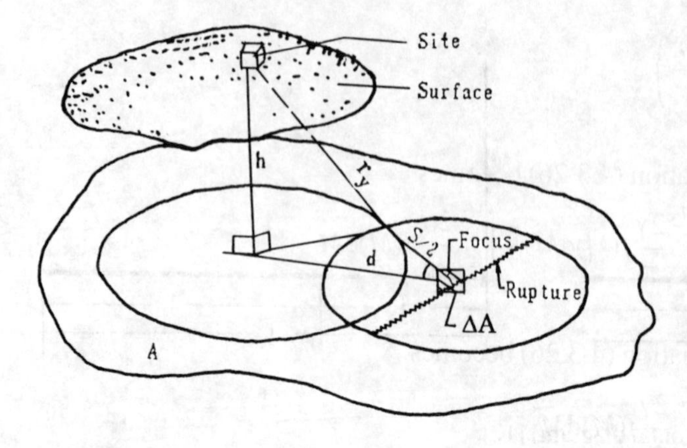

Figure 4.3.3. Mechanism of Source III [12].

1. If $M_1 < M_2 < M_3 < M_u$

$$P\left[Y > y \middle| E_j\right] = \int_{M_1}^{M_2} \frac{2\gamma}{\pi} f(M)\mathrm{d}M + \int_{M_2}^{M_3} \frac{2\alpha}{\pi} f(M)\mathrm{d}M + \int_{M_3}^{M_u} f(M)\mathrm{d}M \tag{4.3.26}$$

In which,

$$M_1 = \max\{M_0, M_1', M_1''\}$$

$$M_1' = g_1(y, h)$$

and M_1'' is the magnitude when $\sqrt{r_y^2 - h^2} + s/2 = d$, and can be determined by

$$\sqrt{(g_2(y, M_1''))^2 - h^2} + \tfrac{1}{2}\exp(aM_1'' - b) = \mathrm{d}$$

M_2 is the magnitude when

$$\sqrt{\left(\sqrt{r_y^2 - h^2}\right)^2 + \frac{s^2}{4}} = d$$

and can be determined by

$$\sqrt{(g_2(y, M_2))^2 - h^2 + \frac{1}{4}\exp(2aM_2 - 2b)} = d$$

M_3 is the magnitude when $\sqrt{r_y^2 - h^2} = d$, and can be determined by

$$M_3 = g_1(y, \sqrt{d^2 + h^2})$$

γ and α can be determined by

$$\gamma = \cos^{-1}\left(\frac{4h^2 + s^2 + 4d^2 - 4r_y^2}{4sd}\right)$$

$$\alpha = \sin^{-1}\left(\frac{\sqrt{r_y^2 - h^2}}{d}\right)$$

2. If $M_2 < M_u < M_3$, Equation (4.3.26) becomes

$$P[Y > y \mid E_j] = \int_{M_1}^{M_2} \frac{2\gamma}{\pi} f(M)\mathrm{d}M + \int_{M_2}^{M_u} \frac{2\alpha}{\pi} f(M)\mathrm{d}M \tag{4.3.27}$$

3. If $M_1 < M_u < M_2$, Equation (4.3.26) becomes

$$P[Y > y \mid E_j] = \int_{M_1}^{M_u} \frac{2\gamma}{\pi} f(M)\mathrm{d}M \tag{4.3.28}$$

4.3.3 *Influence of factor of uncertainty on seismic hazard analysis*

Among factors affecting seismic hazard analysis, the uncertainty of seismicity attenuation is the most significant. Therefore, in calculating the probability of exceedance, the attenuation correction is normally considered.

The attenuation can be assumed to take the following form:

$$Y = b_1 \exp(b_2 M)(f(R))^{-b_3} \tag{4.3.29}$$

Let N_1, N_2, N_3 represent the scattering of data, uncertainty of magnitude and uncertainty of source distance, respectively. Using the coefficients of variation as a designation of uncertainty and introducing correlation factors, one can rewrite the attenuation Equation (4.3.29) as:

$$\begin{aligned} Y_c &= N_1 (N_2 (b_1 \exp(b_2 M)) (N_3 f(R))^{-b_3}) \\ &= N_1 N_2 N_3^{-b_3} (b_1 \exp(b_2 M)(f(R))^{-b_3}) \\ &= NY \end{aligned} \tag{4.3.30}$$

where $N = Y_c / Y$ and let $X = \ln N$, one obtains

$$\ln N = \ln N_1 + \ln N_2 - b_3 \ln N_3$$

Assuming: (1) $\ln N_1$, $\ln N_2$, $\ln N_3$ are random variables with mean values of zero and variation coefficients of δ_1, δ_2, δ_3 respectively; (2) N_1, N_2, N_3 are independent of one another; (3) only the first order approximation is considered, one obtains

$$E(x) = E(\ln N) = E\left(\ln\frac{Y_c}{Y}\right) = \mathrm{E}(\ln N_1 + \ln N_2 - b_3 \ln N_3) = 0 + 0 + 0 = 0$$

$$V(x) = \sigma = \delta_1^2 + \delta_2^2 + b_3^2\, \delta_3^2$$

The probability of exceedance for a given value y can then be expressed as

$$P[Y_c > y] = P[\ln Y_c > \ln y] = \int_{-\infty}^{\infty} p(Y > ye^{-x}) f_x(x) \mathrm{d}x \qquad (4.3.31)$$

Assuming $f_x(x)$ follows the normal distribution,

$$f_x(x) = \frac{1}{\sqrt{2\pi}} \exp\left[-\frac{1}{2}\left(\frac{x-\bar{x}}{\sigma}\right)^2\right]$$

then

$$P(Y_c > y) = \frac{1}{\sqrt{2\pi}} \int_{\infty}^{\infty} p(Y > ye^{-x}) \exp\left[-\frac{1}{2}\left(\frac{x-\bar{x}}{\sigma}\right)^2\right] \mathrm{d}x \qquad (4.3.32)$$

4.4 PROTECTIVE MEASURES AGAINST SEISMIC HAZARDS FOR SITING IN DOCUMENTED AREAS

The foregoing sections provide a theoretical treatment on seismic hazard analysis for siting in earthquake zones with little information. For siting in a well-documented area, however, the process can be simplified and is more direct to seismic design. A typical example is the updated stipulations in the General Provisions of the NEHRP Recommended Provisions for the Development of Seismic Regulations for New Buildings prepared by BSSC (1991) and issued by FEMA, USA in 1992, from which the following sections (4.4.1-4.4.4) are cited.

4.4.1 *Specific criteria in seismic hazards application*

1. *Seismic performance*
Seismic performance is a measure of the degree of protection provided for the public and building occupants against the potential hazards resulting from the effects of earthquake ground motions on buildings, as defined by the provision.

2. *Seismic hazard exposure group*
This is a classification assigned to a building based on its use. This criterion is used in connection with the level of seismicity in assigning buildings to Seismic Performance Categories.

3. *Level of seismicity*
The level of seismicity in well documented area can be represented by seismic ground acceleration maps. The NEHRP Recommended Provision provides Map 1 and Map 2 from which the effective peak acceleration A_a and the effective peak velocity-related acceleration A_v can be determined respectively. Where site-specific ground motions are used or required, they shall be developed with 90% probability of the ground motion not being exceeded in 50 years.

The appropriate location of the building site can be determined on the Maps

Table 4.4.1. Coefficients A_a and A_v (NEHRP Provisions 1991).

Map area from Map 1 (A_a) or Map 2(A_v)	Value of A_a and A_v
7	0.40
6	0.30
5	0.20
4	0.15
3	0.10
2	0.05
1	<0.05*

* For equations or expressions incorporating the terms A_a or A_v, a value of 0.05 shall be used.

and then the values for A_a and A_v can be evaluated from either the maps or Table 4.4.1.

4.4.2 *Seismic hazard exposure group*

The Recommended Provisions specify that all buildings shall be assigned to one of the following seismic hazard exposure groups:

1. Group III: Seismic Hazard Exposure Group III buildings are those having essential facilities that are required for post-earthquake recovery including:

– Fire or rescue and police stations;

– Hospitals or other medical facilities having surgery or emergency treatment facilities;

– Emergency preparedness centres including the equipment therein;

– Power generating stations or other utilities required as emergency back-up facilities for Seismic Hazard Exposure Group III facilities;

– Emergency vehicle garages;

– Communication centres;

– Building containing sufficient quantities of toxic or explosive substances deemed to be dangerous to the public if released.

2. Group II: Seismic Hazard Exposure Group II buildings are those that have substantial public hazard due to occupancy or use including:

– Covered structures whose primary occupancy is public assembly with a capacity greater than 300 persons;

– Buildings for schools through secondary or day-care centres with a capacity greater than 250 students;

– Buildings for colleges or adult education schools with a capacity greater than 500 students;

Table 4.4.2. Seismic performance category (NEHRP Provisions 1991).

Value of A_v	Seismic hazard exposure group		
	I	II	III
$A_v < 0.05$	A	A	A
$0.05 \leq A_v < 0.10$	B	B	C
$0.10 \leq A_v < 0.15$	C	C	C
$0.15 \leq A_v < 0.20$	C	D	D
$0.20 \leq A_v$	D	D	E

– Medical facilities with 50 or more resident incapacitated patients but not having surgery or emergency treatment facilities;
– Jails and detention facilities;
– All structures with an occupancy greater than 5000 persons;
– Power generating stations and other public utility facilities not included in Seismic Hazard Exposure Group III and required for continued operation.

3. Group I: Seismic Hazard Exposure Group I buildings are those not assigned to Seismic Hazard Exposure Group III or Group II.

4.4.3 *Seismic performance category*

A classification for buildings according to the seismicity level of the site and the Seismic Hazard Exposure Group is shown in Table 4.4.2.

In the Recommended Provisions the Seismic Performance Category E is assigned to provide the highest level of design performance criteria. There is a Site Limitation associated with this Category, namely, a building assigned to this Category shall not be sited where there is potential for an active fault to cause rupture of the ground surface at the building.

4.4.4 *Quality assurance*

The Recommended Provisions also provide minimum requirements for quality assurance for designated seismic systems. These requirements are in addition to the testing and inspection requirements contained in the reference standards given in the Provisions. The quality assurance provisions apply to the following:

1. The seismic force resisting system in buildings assigned to category C, D, or E.
2. Other designated seismic systems in buildings assigned to category E.
3. All other buildings when required by the regulatory agency.

CHAPTER 5

Evaluation of seismic parameters of a site

5.1 CLASSIFICATION OF SITE

The response spectrum theory has been used world-wide and adopted by many national seismic design codes. Various site conditions will render different response spectra. The classification of sites may ensure the appropriate selection of response spectrum.

5.1.1 *Criteria of site classification*

Site classification should follow the following principles:

1. Site classification should be directed to its influence on the intensiveness of ground motion and the characteristics of the spectrum.
2. Simplified classification is preferable especially in view of working with insufficient information.

5.1.2 *Overlying stratum*

In engineering geology, the overlying stratum is defined as the loose deposit above the bedrock. However, in considering the ground motion induced by earthquake from the bedrock to the ground surface, the overlying stratum can be defined with particular attention to the change of its shear wave velocity with depth.

Currently, there are two ways of defining the overlying stratum:

1. *With respect to absolute thickness*

The first way is to define the overlying stratum as the entire layer of soil from ground surface down to bedrock. However, from the seismic point of view there are still different understandings about bedrock. In some countries, practical experience suggests that the stratum with a shear wave velocity V_s > 500 m/s can be considered as equivalent to bedrock above which all the soil layers as a whole are recognized as the overlying stratum.

2. *With respect to relative thickness*
The second definition is based on the ratio of shear wave velocity of the lower stratum (V_{sl}) to that of the upper stratum (V_{su}). When this ratio exceeds a certain value, the upper stratum can be treated as the overlying stratum. As a general reference, the following ratio is suggested:

$$\frac{V_{sl}}{V_{su}} \geq 5$$

5.1.3 *Methods of site classification*

There are tens of site classification methods currently used by different designers according to their experience. Two of them are described here as they are relatively independent of local experience.

1. *Soil rigidity method*
Two parameters (the mean shear modulus (G) and the thickness of overlying stratum (h)) are suggested by the Chinese Aseismic Design Code for Structures (1989 Draft) for classifying sites. The mean shear modulus can be obtained by the following equation:

$$G = \frac{\sum_{i=1}^{n} h_i \rho_i V_{si}^2}{\sum_{i=1}^{n} h_i} \tag{5.1.1}$$

where h_i is the thickness of layer i; ρ_i is its mass density; V_{si} is its shear wave velocity; n is the number of layers of the overlying stratum. Equation (5.1.1) is valid for thickness of overlying stratum (h) equal to or less than 20 m.

If the thickness of overlying stratum $h > 20$ m, the average value of shear modulus of the strata within top 20 m is adopted. If the thickness is less than 20 m, the average shear modulus of the total actual thickness is used instead.

Based on both G and h, the site index μ can be determined by Equation (5.1.2):

$$\mu = 0.6\,\mu_G + 0.4\,\mu_h \tag{5.1.2}$$

where μ_G is the contribution of average shear modulus to the site index, as expressed by Equation (5.1.3).

$$\mu_G = \begin{cases} 1 - \exp\left[-0.66\,(G - 30\,000) \times 10^{-5}\right], & \text{if } G > 30\,000 \text{ kPa;} \\ 0, & \text{otherwise} \end{cases} \tag{5.1.3}$$

μ_h is the contribution of overlying stratum thickness to the site index, as expressed by Equation (5.1.4).

Table 5.1.1. Classification of sites.

Type of site	Stiff	Medium stiff	Medium soft	Soft
Site Index (μ)	$1 > \mu > 0.9$	$0.9 > \mu > 0.3$	$0.3 > \mu > 0.1$	$0.1 > \mu > 0$

Table 5.1.2. Classification of construction sites.

Type of ground	Thickness of overlying stratum d_{ov} (m)				
	0	0-3	3-9	9-80	>80
Stiff	I				
Medium stiff		I	I	II	I
Medium soft		I	II		III
Soft		I	II	III	IV

Table 5.1.3. Ground classification according to shear wave velocity.

Type of ground soil	Shear wave velocity V_s (m/s)
Stiff	>500
Medium stiff	250-500
Medium soft	140-250
Soft	<140

$$\mu_h = \begin{cases} \exp(-0.916\,(H-5)^2 \times 10^{-2}) & \\ 0, & \text{for } h > 80 \text{ m}; \\ 1, & \text{for } h \leq 5 \text{ m} \end{cases} \tag{5.1.4}$$

$$\mu_h = \mu_G = 1, \quad \text{for } G > 5 \times 10^5 \text{ kPa or } h \leq 5 \text{ m} \tag{5.1.5}$$

The classification of site based on site index is shown in Table (5.1.1).

2. *Shear wave velocity method*

The Chinese National 'Aseismic Design Code for Buildings' (1989) [39] proposes that the classification of construction site is based on both the classification of ground soil (according to shear wave velocity V_s) and the thickness of overlying stratum as shown in Tables 5.1.2 and 5.1.3.

d_{ov} is the thickness of overlying stratum from the ground surface down to the stiff layer with $V_s > 500$ m/s.

5.2 DETERMINATION OF DESIGN EARTHQUAKE PARAMETERS

5.2.1 *Intensity of ground motion*

Several conventional methods of determining design ground motion parameters are listed as follows:

1. Use seismic hazard analysis result.
2. Convert the given earthquake intensity into ground motion parameters, based on empirical relationship between intensity and ground motion.
3. Evaluate the ground motion parameters from a documented area and use the appropriate attenuation law to determine the parameters of the site being considered.
4. Select parameters according to the seismic zonation.

5.2.2 *Determination of design spectrum*

1. *Statistical method*

The response spectrum is normally derived from actual strong motion record on the site. However, it is very difficult to get appropriate sample in a limited period of available historical records. An alternative method is to use an actual strong record obtained on a site of similar geological and lithological conditions by the following procedures:

a) Collect and select strong accelerograms of similar sites.

b) Normalize the accelerograms selected.

c) Calculate the response spectra of each accelerogram with different damping ratios.

d) Classify the calculated spectra and develop the design spectrum by simulation; in fact, all the response spectra adopted in various national codes are results of statistical studies.

2. *Seismic hazard analysis*

A response spectrum, corresponding to a probability of exceedance specified by the governing code can be used as the design spectrum with consideration of the following factors:

a) The influence of all the potential seismic sources; and

b) The influence of the major seismic sources.

5.3 ACCELERATION-TIME HISTORY FOR SEISMIC DESIGN

The following key points should be particularly considered in siting for seismic design.

5.3.1 *The modes of ground motions*

All the seismic records can be classified into three categories:

1. Single cycle shaking which normally denotes minor shock with very shallow focus and limited damage around epicentre, e.g. Hueneme port earthquake (1957) (Fig. 5.3.1).

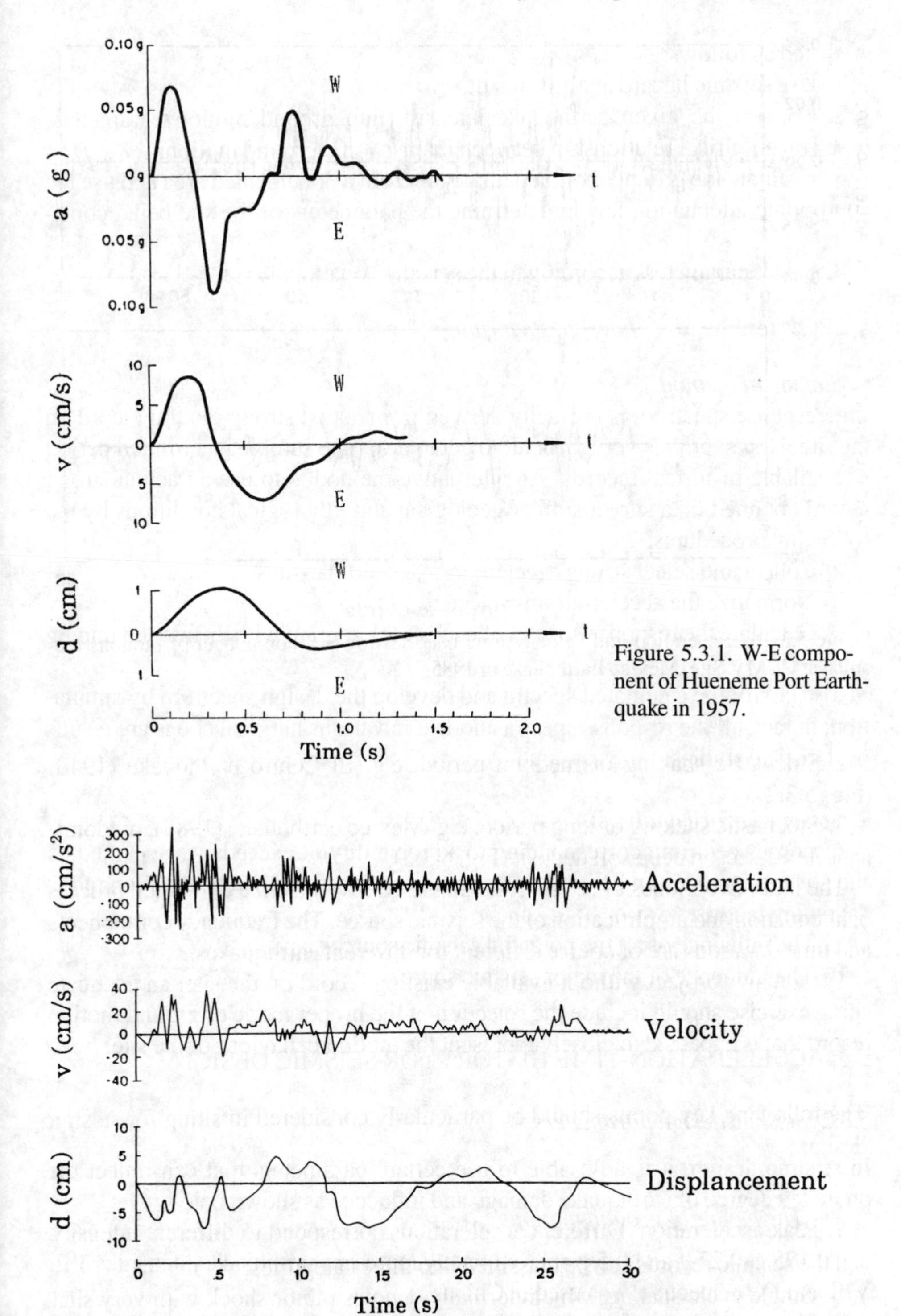

Figure 5.3.1. W-E component of Hueneme Port Earthquake in 1957.

Figure 5.3.2. N-S component of California El Centro Earthquake in 1940.

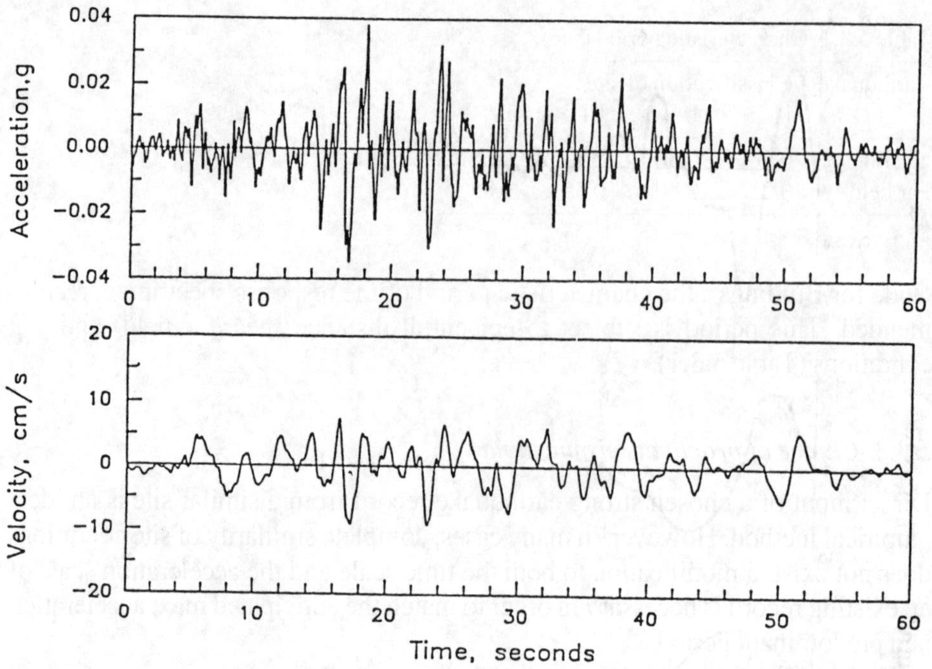

Figure 5.3.3. Acceleration and velocity time histories for N-S component of motions recorded at CUMV Site, Mexico Earthquake in 1985.

2. Stochastic shaking of medium period, e.g. El Centro earthquake (1940) (Fig. 5.3.2).

3. Stochastic shaking of long period, e.g. Mexico earthquake (1985), predominant $T \simeq 2$ sec on deep soil deposit (Fig. 5.3.3).

The latter two modes of seismic shaking are derived from a combined process of attenuation and amplification of the seismic source. The frequency components and time durations are of course different for different earthquakes.

For a major project without available existing record on the site, an adequate siting exercise should include the selection of the proper mode of ground motion record that is expected to closely represent the motion behaviour of the site.

5.3.2 *Seismic design parameters*

In seismic design, it is advisable to use certain parameters that can reflect the predicted degree of earthquake damage and influence as shown below.

1. Peak acceleration. Different accelerations correspond to different intensity, e.g. 0.125 g, 0.25 g and 0.5 g approximately represent earthquake intensities VII, VIII and IX, respectively, as used in China.

2. The mode-shape of structure. The mode shape reflects the site condition and the seismic influence. For example, in Chinese National Code 'Aseismic Design

Table 5.3.1. Characteristic period (sec).

Earthquake	Classification of site			
	I	II	III	IV
Near field	0.20	0.30	0.4	0.65
Far field	0.25	0.40	0.55	0.85

Code for Buidings', the characteristic period of the response spectrum is recommended. This period is subject to epicentral distance (near/far field) and soil conditions (Table 5.3.1).

5.3.3 *Choice of practical seismic input*

Direct input of a chosen strong earthquake record from a similar site is an ideal empirical method. However, in many cases, complete similarity of site conditions does not exist, a modification to both the time scale and the acceleration scale of an existing record is necessary in order to match the anticipated max. acceleration and predominant period.

The modification process is consisted of:

1. Evaluate the maximum acceleration (a_{max}), predominant period (T_0) and time duration (t_d) associated with the predicted earthquake magnitude, epicentral distance and site conditions.

2. Choose a strong seismic record whose maximum acceleration (a'_{max}) and predominant period (T'_0) are approximately equal to (a_{max}) and (T_0), and the ratio of duration (t'_d/t_d) is also close to or smaller than (T'_0/T_0) respectively.

3. Obtain modified acceleration-time history by multiplying the coordinates of the chosen strong earthquake motion record with the ratio a_{max}/a'_{max}, T_0/T'_0 respectively.

CHAPTER 6

Seismic effect of fault and faulting

6.1 DEFINITION AND CLASSIFICATION OF FAULT AND FAULTING

Fault is a general term covering all the tectonic ruptures in rock formation caused by tectonic movement in the crust and the upper mantel. Faulting means tectonic rupture in the overburden layer by either ground shaking or the extension of the fault dislocation of the bedrock. In this book faulting may also mean the process of such rupturing. When the dislocation reaches the ground surface, it is called surface faulting.

Fault can be classified into two categories: active fault and non-active fault. The former is of major interest to earthquake engineering. While the latter may have influence on the propagation of seismic waves in the phenomena of attenuation, reflection, refraction and focusing.

6.1.1 *Active fault*

1. *Definition*

There has been no unified definition world-wide; the major divergence of views lies in both the duration and the manner of being active.

From the geological point of view, a tectonic rupture that has been active since the quarternary period and likely to be active in the future can be considered as an active fault.

From the engineering point of view, as suggested in this book, a fault is considered active only if there has been some activity since the upper holocene epoch (ten thousand years ago) and may be active within the next one hundred years.

2. *Mode of activity*

There are three modes of activity: shaking (shock), creeping and dislocation which occurs suddenly. From the traditional geological view point, a fault is considered active if any such mode of activity has occurred within the time period as defined above.

In earthquake engineering, it is contended here that only the shaking (vibration) can be designed for. Creeping is essentially a matter of non-accelerated motion and has nothing to do with shock resistance. As for dislocation (except surface faulting in soil), the relative displacement is generally too large and too strong to resist. It is beyond the capability of seismic design and is preferable to avoid sites of such dislocations.

6.1.2 *Causative fault and capable fault*

1. A causative fault is a tectonic fracture zone where either a historical or a forecasted seismic focus is located.

In geotechnical earthquake engineering, only events of magnitude $M > 5$ are of concern because of their possible destructive nature, whereas in seismo-geology, events of $M > 3$ can be related to causative fault.

2. The concept of capable fault has been defined by the US Atomic Energy Committee. This type of fault is directed to nuclear power plant protection and generally not applicable to other construction projects.

6.1.3 *Non-active fault*

A non-active fault is one formed in ancient geological history, has not been active since the upper holocene epoch (recent ten thousand years) and is unlikely to be active during the life of the engineering project. According to Chinese Code of Geotechnical Investigation (1989 draft), aseismic consideration is not required for structures located in zones with non-active faults. It is, however, advisable to design for differential settlement for structures founded on highly fractured rocks in regions with non-active faults.

6.1.4 *Tectonic rupture in soil*

Tectonic rupture in soils is a ground rupture caused by seismic ground waving. The mechanism is 'tectonically' identical to the mechanism of a seismic source, i.e. the causative fault. However, it is not directly linked to the causative fault.

A tectonic rupture in soil may reach a maximum length of several kilometres on the ground surface and tends to decrease rapidly with depth. It normally vanishes within several metres below the surface.

6.2 CAUSATIVE FAULT IN SEISMIC HAZARD ZONING

Since disastrous earthquakes are definitely related to seismo-geotectonic background, the zonation of seismic hazard should begin with the identification of a causative fault followed by the study of periodic recurrence and migrating character of its activity.

6.2.1 *Identification of causative fault*

The following rules are suggested for identifying causative faults:

1. Causative faults are affiliated with active fault zones.
2. The maximum magnitude of an earthquake generated from the causative fault is a function of the depth of the fault dislocation, which can be estimated according to the mechanism of different fault structures.
3. Not all faults along a big abysmal fault are causative. Only the part with geotectonic stress concentration is likely to be causative, such as:

- Ends of the fault;
- Bending points of the fault;
- Fault intersections;
- Strong differential movement zones of the fault block in the quarternary system;
- A zone with numerous destructive earthquakes since the holocene epoch.

6.2.2 *Fault length versus magnitude*

The relationship between the length (L) of a causative fault and the magnitude of earthquake generated from the fault varies with districts of different geotectonic background. Generally speaking, the relationship obeys a semi-logarithmic law and takes different forms according to different authors. Some examples are given in the following:

Selemon (1982)

$$M = 0.809 + 1.34 \lg L \text{ (normal fault)}$$
$$M = 2.02 + 1.14 \lg L \text{ (thrust fault)}$$
$$M = 1.40 + 1.17 \lg L \text{ (strike-slip fault)}$$

D.S. Chen (1984)

$$M = 6.43 + 0.665 \lg L \text{ (Western part, China)}$$
$$M = 6.64 + 0.565 \lg L \text{ (Eastern part, China)}$$
$$M = 6.72 + 0.482 \lg L \text{ (Taiwan, China)}$$

Widsman & Major (1969)

$$M = 3.3 + 1.7 \lg L$$

Wang et al. (1983) [35]

$$M = 6.22 + 0.635 \lg L \text{ (Worldwide)}$$

The above listed equations can be used to estimate either the fault length or the magnitude for regions with similar seismicity as those indicated with the equations.

In addition, if dislocation is also considered, the following relationship by Ambrasseys et al. (1984) can be used:

$$M = 1.1 + 0.4 \lg (L^{1.58} D_d^2)$$

where D_d is the dislocation of fault in cm.

Generally speaking, all these empirical relationships are applicable to earthquakes of medium to shallow focal depth, i.e. $h = 10 \sim 20$ km, and $M \geq 6.0$. However, certain statistical errors were inevitably introduced into the historical data due to the confusion in identifying tectonic ruptures from causative faults, and sometimes due to the difficulty of measuring the length of the invisible fault. The above regression equations should therefore be used with caution and in association with knowledge of local seismo-geological background.

6.3 INFLUENCE OF CAUSATIVE FAULT ON EARTHQUAKE INTENSITY

6.3.1 *Earthquake intensity along the causative fault zone*

There are two aspects to the influence of causative fault on ground damage:

1. *Direct effect*
Ground shaking and surface faulting acting directly on buildings and ground facilities are the direct consequences caused by the elastic rebound of a causative fault dislocation.

2. *Indirect effect*
Soil liquefaction, landslide, land subsidence, ground settlement and rockfall, etc. as induced by ground shaking generated from the causative fault are the secondary effects which form the major components of earthquake intensity evaluation.

Generally speaking, earthquake intensity will be higher along the causative fault zone than in areas away from it. However, in considering ground motion, dislocation, liquefaction, landslide etc., the basic intensity specified for seismic design with respect to a certain district does not need to be modified as a result of the proximity to a causative fault. This is because effects of the causative fault should have been considered in the regional basic intensity determination.

It is necessary to point out that local damage near the causative fault zone appearing on ground surface is not necessarily heavier, or may even be lighter sometimes. This phenomenon is caused by the fact that the occurrence of the surface rupture or faulting absorbs local ground shaking energy, shortens shaking duration, and consequently decreases damage to buildings adjacent to the faulting zone.

It should be noted that the ground dislocation as the extension of causative fault to ground surface is too overwhelming to resist. However, its particular location can be investigated during siting with the aid of surface geological survey and digging exploration trenches. Then, one can plan the proposed building and ground facilities away from the faulting zone. It has been shown that the tectonic surface faulting or the extension of causative fault keep recurring at the same location (Wang 1981, 1983).

6.3.2 *Earthquake intensity at the intersection of causative fault and non-causative fault*

Historical evidence shows that intensity increases at the intersection of a causative fault and a non-causative fault if the faults are large, but intensity does not change if the faults are small.

6.3.3 *Earthquake intensity along non-causative faults*

No difference in earthquake intensity was found between areas along and away from non-causative faults. However, an abysmal non-causative fault has some attenuation or shielding effect on seismic waves.

6.4 SURFACE FAULTING

Surface faulting is a visible dislocation on the ground surface as a result of a strong earthquake. In earthquake engineering, this term normally covers all the ground ruptures occurred in the overlying soil or the outcrop caused by either intensified and tectonically orientated ground movement or by extension of an apparently shallow and embedded causative fault. Historical records show that earthquakes with a magnitude $M \geq 6$ might induce surface faulting. This is true not only for plate-boundary earthquakes but also for intra-plate earthquakes (Adams et al. 1991). Therefore, surface faulting is one of the inevitable seismic hazards during a strong event.

6.4.1 *Recurrence of surface faulting*

It has been well documented that most surface faultings occur repeatedly in the same area, particularly along a fault belt where dislocations happened in the quarternary period. This is due to the fact that it is easier for seismic energy to be released along structural weak zones or fault zones than in the surrounding intact rock mass.

For an underlying strike slip fault, the location and strike of the surface faulting are basically identical to previous ones irrespective of whether or not overburden exists (Fig. 6.4.1A).

For an underlying reverse (thrust) fault, the surface faulting recurs with the

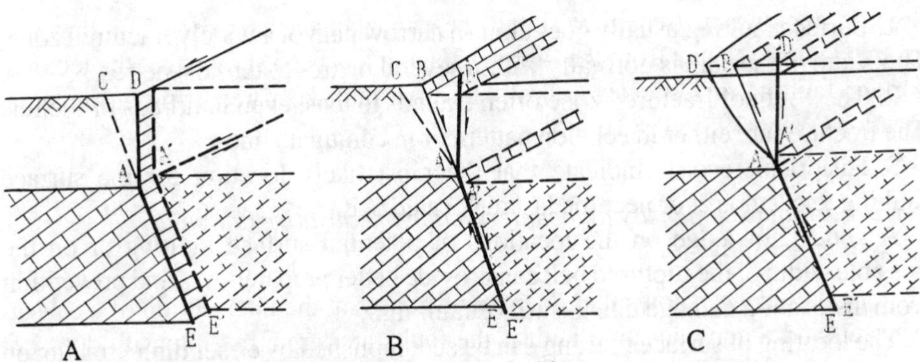

Figure 6.4.1. Influence of overburden on surface faulting.
A. Strike-slip fault, B. Normal fault, C. Reverse fault.
AE-A′E′ – Action of causative fault;
AB, AD, AD′ – Width of the occurrence of possible surface faulting;
AC – Impossible surface faulting;
AA′, EE′ – Fault displacement.

same strike, but not necessarily at the same location. Most likely, the location will be shifted (Fig. 6.4.1C).

For an underlying normal (thrown) fault, the surface faulting recurs and is located in between the previous ones with the same strike (Fig. 6.4.1B).

The following factors are closely related to the occurrence of surface faulting:

1. Creeping of a fault on the outcrop may result in surface faulting. When the overburden soil is sufficiently thick(> 5 m), surface faulting is unlikely to occur.

2. Sudden dislocation of a shallow embedded causative fault may produce surface faulting.

3. For strong earthquakes with a magnitude $M \geq 6$ from a shallow source (focus depth 10 ~ 30 km) in the crust, causative fault is able to produce a surface faulting. When $M \geq 7.2$, surface faulting (including tectonic surface rupture in soil) is almost inevitable (Wang et al. 1983) [35].

4. The maximum thickness of the overburden soil is about 30 m for surface faulting to occur as an extension of the dislocated causative fault. More details are given in a later section.

The Tangshan Earthquake (1976, $M = 7.8$) is a good example illustrating that surface faultings behave as tectonic ruptures and occur at the same location as a set of older ruptures in the soil layers at a number of different sites (Wang 1983).

It has been widely recognized in the west coast of the United States that surface faulting in well documented district like California State can be more easily delineated than other types of earthquake damages (Reitherman 1991).

6.4.2 *Possible size of surface faulting*

Site investigation in strong earthquake zones has established the following rules:

1. Surface faulting usually goes along a narrow path of a widely fractured zone. The width of the path is normally within several metres to tens of metres.
2. The width of fractured zone often extends to tens, even hundreds of metres. The fractures are either in echelon pattern or in conjugate lines.
3. Historical records indicate that the most likely location for the surface faulting to occur is at or near the last dislocated belt.

A good knowledge on the locations of potential surface faulting or on the possible width of the ruptured belt can provide better preparedness and prevention from dislocation damage due to surface faulting.

The locating of surface faulting can be accomplished by either direct tracing on ground surface, digging trenches across a suspected faulting, or running a close-up trenching at the proposed construction site.

6.4.3 *Dislocation of surface faulting*

The amount of dislocation (D) of a potential surface faulting can be estimated with the D versus M relationship, or by referring to the records of historical surface faulting of a well documented area. Table 6.4.1 shows some statistical relationship between D and M based on past records.

It should be noted that, the regression equations in Table 6.4.1 are associated with certain statistical scatter. These $D \sim M$ relationships should therefore be used only for preliminary estimates.

6.4.4 *Influence of overburden on surface faulting*

The thickness of overburden has important influence on the appearing of surface faulting because the overburden has ability to absorb dislocation originated from the underlying rock mass. The occurrence of surface faulting therefore depends largely upon the total thickness and the critical strain of the overburden soil.

The required overburden thickness to prevent surface faulting is dependent on the type of causative fault dislocation. For the strike slip causative fault dislocation, it does not need a thick overburden. For the normal and reverse faults, greater overburden soil thickness is required to suppress surface faulting. In the case of

Table 6.4.1. Dislocation D (m) versus M relationship.

Equation and reference	Source of Data
$\log D = 0.55M - 3.71$ (Iida, 1965)	Worldwide
$\log D = 0.96M - 6.69$ (Chinnery, 1969)	Worldwide
$\log D = 0.57M - 3.91$ (Bonilla, 1970)	USA
$\log D = 0.6M - 4.0$ (Matsuda, 1975)	Japan
$\log D_{max} = 0.57M - 3.19$ (Bonilla, 1970)	USA
$\log D_{max} = 0.67M - 4.33$ (Meishi, 1972)	Japan

the reverse fault the critical thickness (t_{cr}) of overburden soil beyond which surface faulting will not occur is given by

$$t_{cr} = \frac{D_f}{\varepsilon_{cr}} \qquad (6.4.1)$$

where ε_{cr} is the critical strain of the overburden soil (5% for stiff soil, 10% for soft soil). D_f is the estimated maximum dislocation of the causative fault in the bedrock.

6.4.5 *Important role of surface faulting in seismicity evaluation*

Surface faulting is the unique trace and evidence of strong earthquake permanently left on the ground. It is visible and measurable either directly or indirectly through some form of exploration, with trenching being preferable. A thorough survey of surface faulting can provide important information on the historical seismicity. It can also improve forecast of the strong events in the following ways:

1. *Amending aspects of historical records*
– The frequency of prehistorical strong earthquakes;
– The time interval of the recurrence of strong historical events;
– The magnitude and intensity of past events;
– The occurrence (strike, dip, relative displacement, width of disturbed zone, etc.) and character of the surface faulting possibly reflecting the seismic source mechanism.

2. *Predicting future events*
– The likelihood of occurrence of the next strong earthquake;
– The possible character and size of the disturbed zone and the potential dislocation of the surface faulting;
– The probable intensity of the future event.

3. *Siting and planning*
Since surface faulting is usually confined to a relatively narrow zone along existing faulting ruptures, it is possible to locate proposed structures away from the existing rupture zones.

Emphasis is needed on the determination of minimum earthquake magnitude which may rupture the ground surface because surface faulting is relatively common during major earthquakes with a shallow focus, but rare or absent during small earthquakes.

6.5 EVALUATION ON ACTIVE FAULT AND COUNTER MEASURES

6.5.1 *Non-causative fault*

The term non-causative fault refers to those active faults experiencing creep and no noticeable shock in the past ten thousand years. The evaluation and counter-measures are mainly directed to the creep problems and local seismic anomaly as shown below:

1. As a general requirement for most buildings, avoid sites with potential creep ruptures. However, for many linear projects, like pipelines, railways, highways, etc. where intersections with the fault displaying creep is inevitable, certain flexible linkage or soft connection should be adopted.

2. Higher intensity anomaly is likely to occur at the intersection of a non-causative fault and a causative fault. The seismic effects at such an intersection should be checked against historical experience.

6.5.2 *Causative fault*

1. For an ordinary project, seismic design can be carried out according to the officially authorized seismic zonation and the associated design earthquake parameters. Proposed structures should not be constructed in causative fault zone which can be identified in advance by seismo-geological survey.

2. For structures of special significance and importance, the following evaluation and measures should be taken:

a) Detailed investigation and decision making. Determine whether or not causative faults as defined in this chapter do exist, and ascertain the seismic hazard for the site as described in Chapter 4.

b) Evaluation of seismic parameters of the site. As suggested in Chapter 5, the classification of site and its ground soils, the design earthquake parameters and the response spectrum should be selected with careful study.

c) Keeping out of a causative fault. All proposed structures of some importance to be constructed should be kept out of the fault zone, especially out of the intersection of two active faults, bending corner of abysmal fault, ends of causative fault, and areas of high tectonic stress concentration.

6.6 GROUND RUPTURE

6.6.1 *Tectonic rupture and its seismic effect*

1. *Character and origin of tectonic rupture*

In a strong earthquake, a series of en echelon cracks on the ground is induced by intensive ground shaking. The cracks are normally orientated as a whole in the same direction as the causative fault. However there is no direct linkage between the cracks and the fault. The dislocation of rupture is largest at the ground surface and tends to diminish with depth in soils. The rupture should by no means be treated as the fault in the bedrock. It should be properly recognized as surface faulting.

2. *Seismic effect of tectonic ruptures*

The formation of tectonic ruptures is the consequence of ground shaking and results in the release of ground motion energy and reduction of duration of ground shaking. The rupture itself is therefore not harmful as a dynamic disturbance of the ground motion, nor can it intensify any ground damage. In any case, ground rupture is a kind of ground failure which will not cause additional seismic damage to the buildings by shaking.

Since tectonic ruptures extend downward to a limited depth (normally several metres only), there will be less or even no cracking influence on deeply embedded chambers, tunnels and other underground facilities.

6.6.2 *Non-tectonic ground rupture*

1. *Characteristics*

Non-tectonic rupture is another kind of ground failure composed mainly of gravitational slide and characterized as follows:

a) Vertical displacement (slide) is always greater than horizontal movement.

b) Mainly tensional cracks at the boundary of land subsidence and settlement or tension-compression coupled cracks (horizontal movement).

c) No obvious orientation among all the ruptures.

2. *Cause and places of non-tectonic rupture*

a) Land subsidence due to earthquake;

b) Gravitational landslide induced by seismically liquefied ground;

c) Locally intensified ground waving resulting in lateral squeezing and diagonal tension.

3. *Treatment*

It should be noted that although the ground failures described in this section are damages resulted from earthquake, they have mechanisms and behaviours different from those of tectonic ruptures. Therefore they should be neither treated as nor confused with surface faulting. They should be dealt with as landslides caused by either liquefaction or by slope failure as illustrated in the following chapters.

CHAPTER 7

Seismic liquefaction of soil

7.1 FUNDAMENTAL CONCEPT OF LIQUEFACTION

7.1.1 *Definition of liquefaction*

The definition of liquefaction is directly linked to the criteria of liquefaction evaluation. Since there has not been any internationally standardised definition of soil liquefaction, different understanding and handling of this problem often arise. In view of such a confused situation, it is necessary to state the authors' point of view in the following:

Seismic liquefaction of soil is essentially a pictorial term to visualize the whole process and consequence of saturated cohesionless/low-cohesion soil turning into a liquid state due to a sudden loss of shear strength under seismic action.

According to the definition of terms related to liquefaction by ASCE (1978) and the authors' experience, the following definitions are suggested for both laboratory and field work:

1. *Liquefaction*
The process or state of saturated granular soils behaving like a liquid as a consequence of increased pore pressure and reduced effective stress.

2. *Sand boil*
An ejection of sand and water caused by piping from a zone of excess pore pressure within a soil mass.

3. *Initial liquefaction*
The state of soil when the pore water pressure first approaches or equals to the total applied pressure under cyclic loading conditions.

4. *Microscopic liquefaction*
The liquefied state of a point in a soil mass as defined by (1) or (3) above, or as

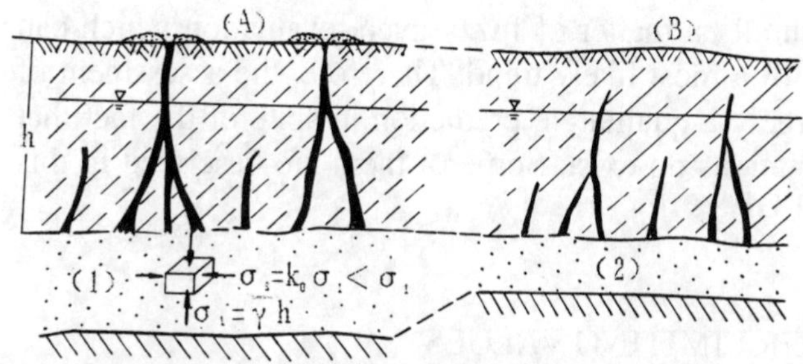

Figure 7.1.1. Different assessment conclusions by different methods – Microscopic/macroscopic.
Microscopically: (1) Liquefied (2) Liquefied
Macroscopically: (A) Liquefied (B) Non-Liquefied

assessed by comparing the dynamic strength of a soil where a sample taken from the point with the dynamic stress estimated for the same point.

5. *Macroscopic liquefaction*
The liquefied state of a soil mass of significant size as assessed by taking into consideration of soil strength, seismic stress and influence of local geology and topography. This type of liquefaction is normally identified by the occurrence of sand boils on level ground.

Macroscopic liquefaction is further illustrated in Figure 7.1.1. This figure shows a real example of a site with varying thicknesses of liquefiable soil layers and resulted in different situations. Because of the same magnitude, epicentral distance, duration of vibration and effective overburden pressure at the top of the sand layer, soil liquefies at both locations (A) and (B). However, sand boils on the surface develops only at the location (A) directly above the thicker layer where excess pore water is sufficient to push the sand particles out of the ground. So macroscopic liquefaction is considered to occur only at this location. In fact, location (B) is usually deemed to be not liquefied, because no evidence or even no damage of liquefaction can be found on the level ground.

Macroscopic liquefaction is so important that it actually provides all the real case records of soil liquefaction on level ground all over the world. These case histories form the only sound and reliable engineering basis for establishing empirical formulations. Without empirical comparison, no liquefaction assessment can be made with practical significance.

7.1.2 *Liquefaction potential*

The internal cause of seismic liquefaction is the soil condition, whereas the external cause is the seismic action which is composed of a number of uncertain factors associated with earthquake events. Due to such uncertainty, the assessment of liquefaction is naturally a matter of fuzzy event evaluation which can only produce a conclusion of a most likely trend. Therefore, the assessment of liquefaction potential is largely a qualitative prediction in spite of the fact that many quantitative determinations do exist. Some of them are described in this chapter.

7.2 FACTORS AND THEIR LIMITING VALUES

Factors that influence seismic liquefaction mainly come from three aspects: seismic action, soil and environment (local topographic features, ground water regime etc.) conditions. From numerous historical records and investigations, the critical indices showing liquefaction can be summarized as follows.

7.2.1 *Threshold value of earthquake intensity causing liquefaction*

In Chinese historical records of seismic events, no liquefaction has ever occurred at a site which experienced earthquake of a magnitude less than 5($M < 5$) with a medium-shallow focal depth and with an intensity roughly less than 6 ($I <$ VI). This is also true and applicable to other parts of the world.

For seismic liquefaction evaluation it is thus appropriate to consider a threshold intensity of VI beyond which liquefaction may occur.

7.2.2 *Maximum epicentral distance*

For a seismic event, there is a maximum epicentral distance beyond which the earthquake effects are no longer strong enough to trigger liquefaction. By summing up hundreds of liquefaction case histories that occurred in the past several hundred years, the following relationship is obtained [38]:

$$\Delta_{max} = 0.82 \times 10^{0.862(M-5)} \tag{7.2.1}$$

where Δ_{max} is the maximum epicentral distance in km and M is the magnitude.

7.2.3 *Maximum depth of liquefiable soil*

Since the effective overburden pressure directly controls the occurrence of liquefaction, there is a maximum depth below which the soil will not liquefy due to high effective overburden pressure or small stress ratio of horizontal seismic shear stress to vertical effective stress.

Observations based on the liquefaction experience on natural level ground in earthquake regions of many countries shows that no liquefaction has occurred below 15 m from the ground surface.

7.2.4 *Maximum depth of groundwater table*

Experience also shows that in most areas suffering liquefaction hazard, the groundwater table is relatively shallow, say, not deeper than 3 m. A few cases occurred with groundwater table 3 ~ 4 m below ground surface. No liquefaction has been observed for level ground sites with a water table deeper than 5 m from the surface.

7.2.5 *Kinds and characteristics of liquefiable soils*

Saturated sand, coarse sand, fine sand, silty sand and even sandy silt can all liquefy when there is insufficient drainage boundary around them. As a general guide, the following indices are characterictics of liquefiable soils:

1. Mean size: $d_{50} = 0.02 \sim 1.00$ mm;
2. Fines ($d \leq 0.005$ mm) content not over 10%;
3. Uniformity coefficient $(d_{60}/ d_{10}) < 10$;
4. Relative density $d_r < 75\%$;
5. Plasticity index $I_p < 10$.

7.3 EVALUATION OF LIQUEFACTION POTENTIAL

7.3.1 *Macroscopic evaluation*

Seismic liquefaction is characterized by three key points:

1. Historical liquefaction occurred repeatedly in the same area, even at the same site.
2. Liquefaction only occurred in areas of basic intensity equal to or greater than VI ($I \geq$ VI).
3. Liquefaction only took place in very recent deposits, say within one or two thousands of year.

The method of assessment based on these characteristics is shown in Figure 7.3.1 for natural level ground.

7.3.2 *Microscopic evaluation*

For assessing liquefaction potential of man-made and/or sloped natural ground, the microscopic method is especially applicable. As a general guide, the method is composed of two parts: testing for soil strength and comparing strength with seismically induced stress. Testing can be conducted in the laboratory or in situ. A schemetic diagram of this procedure is shown in Figure 7.3.2. The method can be simplified as more information becomes available.

For more important projects, both macroscopic and microscopic assessment should be carried out.

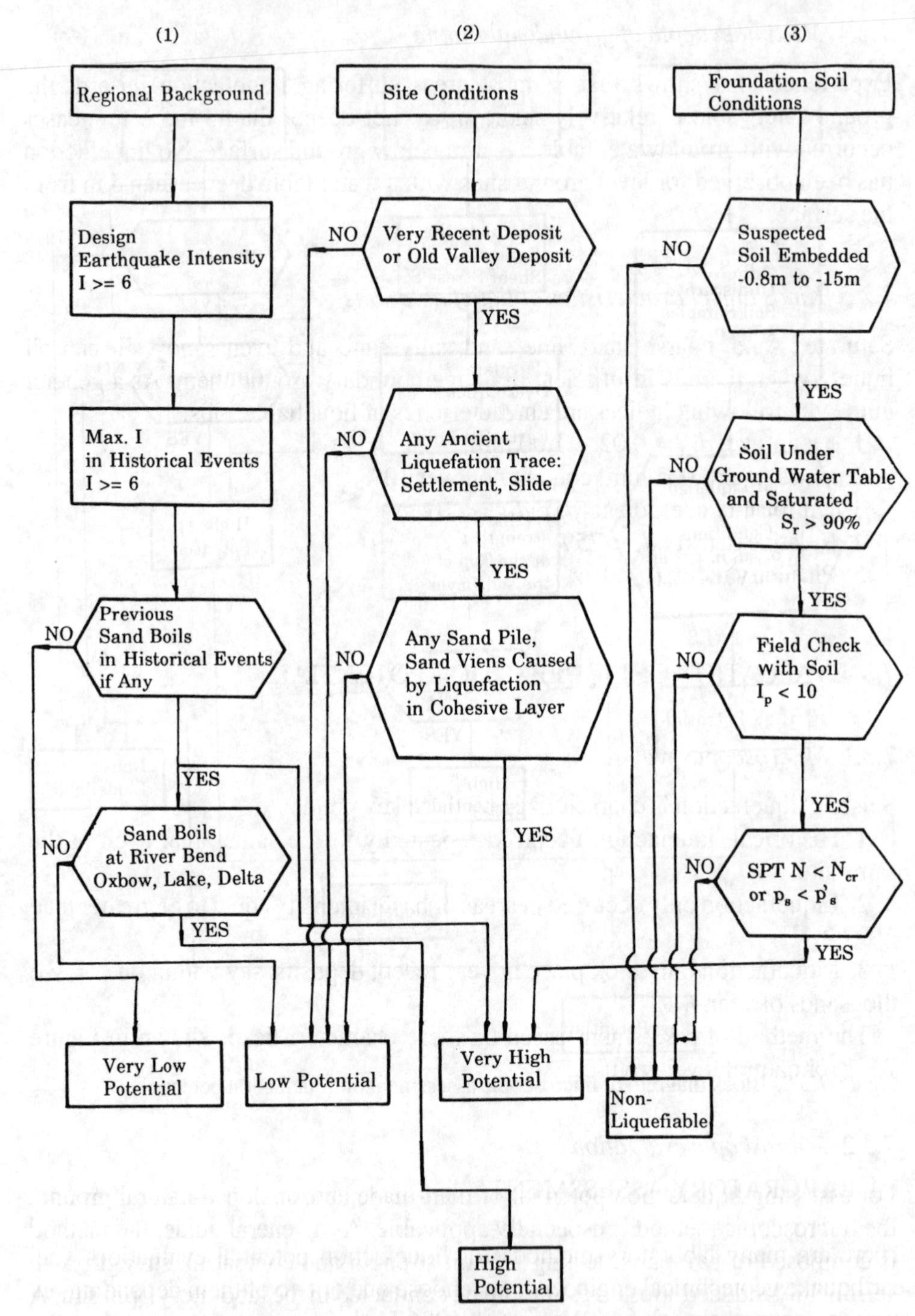

Figure 7.3.1. Block diagram of macroscopic assessment of liquefaction potential.

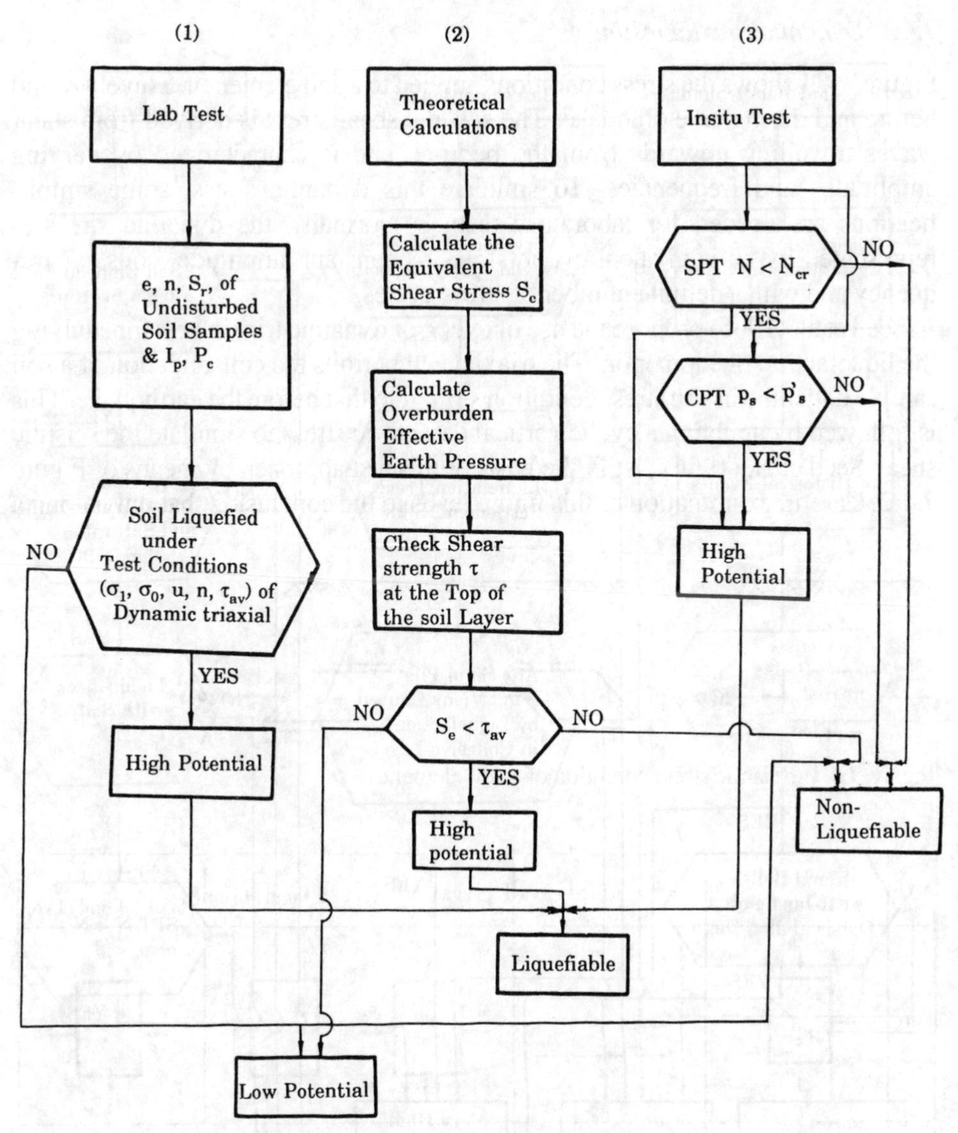

Figure 7.3.2. Block diagram of microscopic assessment of liquefaction potential.

7.4 LABORATORY ASSESSMENT

There are many laboratory methods for liquefaction potential evaluation. The earthquake geotechnical engineers may choose one or two of them depending on requirement and experience (Wang et al. 1986)[13].

7.4.1 *Dynamic triaxial testing*

Figure 7.4.1 shows the stress conditions applied to a soil element in a level ground before and during an earthquake. The seismic shear stress is derived from shear waves travelling upwards from the bedrock and is characterized by varying amplitudes and frequencies. To simulate this dynamic stress, some simplifications are needed for laboratory testing. Normally, the dynamic stress is represented by an equivalent cyclic stress of constant amplitude, constant frequency and with a definite number of stress cycles.

Seed et al. (1966) propose the use of cyclic or dynamic triaxial test for studying the liquefaction phenomenon. The triaxial cell permits the consolidation of a soil sample under the in situ stress condition similar to that before the earthquake. This is followed by applying a cyclic vertical deviatoric stress to simulate the seismic shear. Seed et al. (1966) justify the basis of this test approach by means of Figure 7.4.2. Careful examination of this figure leads to the conclusion that only along a

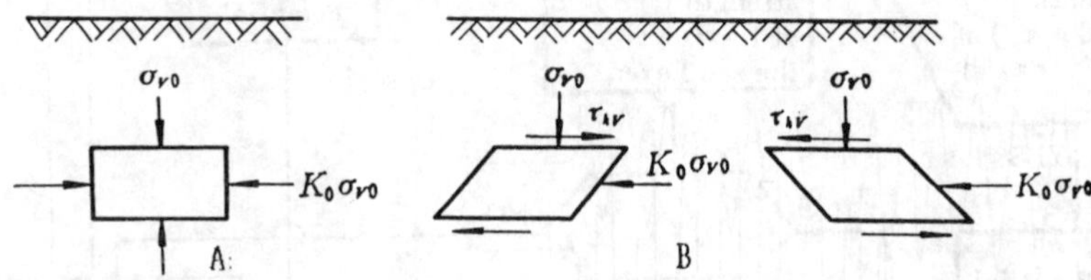

Figure 7.4.1. Seismic stress condition of a soil element.

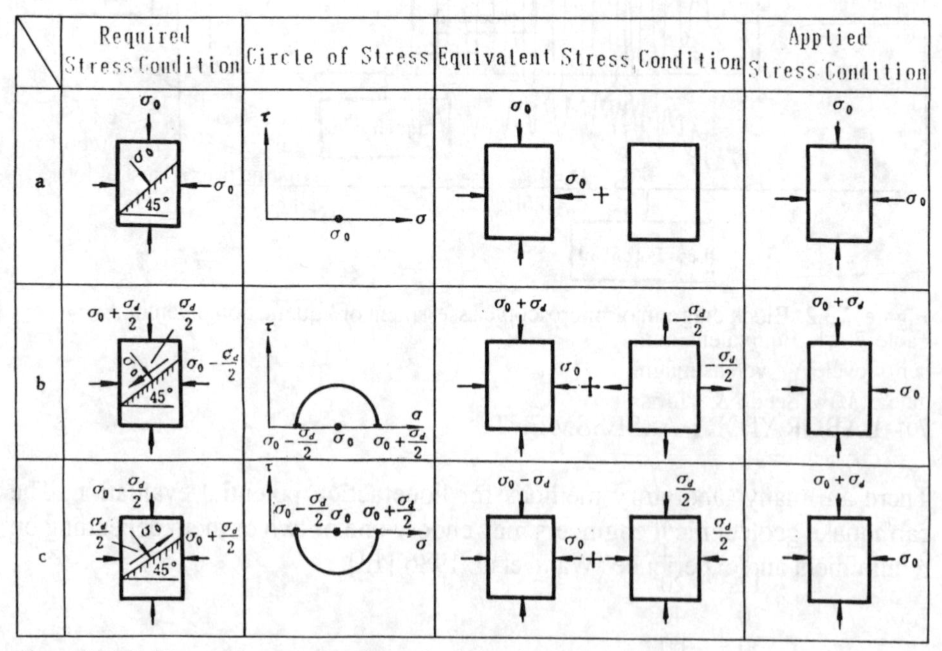

Figure 7.4.2. Dynamic stress simulation in dynamic triaxial test.

plane at 45° from the vertical does the stress condition roughly simulate the reality.

Figure 7.4.3 gives an example of typical test results of a cyclic triaxial test on a saturated loose sand. The liquefaction state of the soil can easily be identified either by considering the point when the pore pressure is equal to the confining pressure or when the peak-to-peak value of dynamic axial strain exeeds a certain limit, say 10%.

The applied cyclic shear stress is taken as the liquefaction strength τ_l corresponding to the number of stress cycles to reach liquefaction (N_{eq}). Table 7.4.1 shows correlations among N_{eq}, earthquake magnitude M and duration t. The information in this table enables the estimation of liquefaction strength for different earthquake magnitudes.

The criterion of liquefaction failure is simply expressed as the condition that the equivalent seismic shear stress (τ_{eq}) induced by an earthquake exceeds the

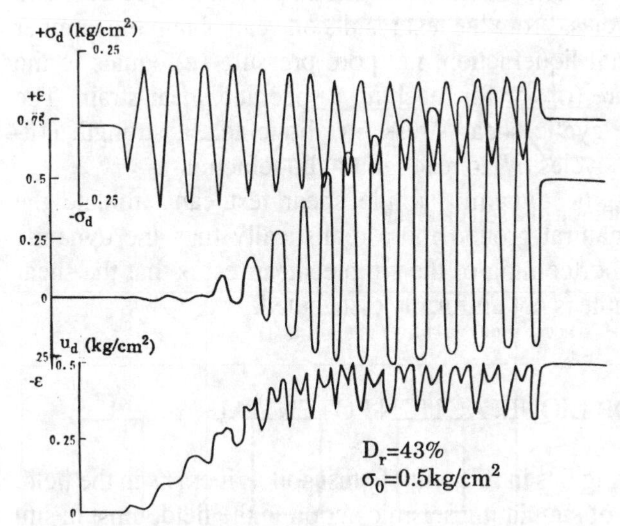

Figure 7.4.3. Typical record of dynamic triaxial test of loose sand.

Table 7.4.1. Equivalent number of cycle N_{eq} versus magnitude (M) (Seed & Idriss, 1971).

M	N_{eq}	$t(s)$
5.5-6.0	5	8
6.5	8	14
7	12	20
7.5	20	40
8	30	60

liquefaction strength (τ_l). Studies based on statistical analyses show that τ_{eq} can be obtained from (Seed & Idriss 1971):

$$\tau_{eq} = 0.65 \frac{a_{max}}{g} \gamma h r_d \qquad (7.4.1)$$

where a_{max} = maximum peak acceleration; g = gravitation acceleration; γ = bulk density of soil; h = depth of soil; r_d = correction coefficient for the depth of soil. According to Shibata & Teparaksa (1988):

$$r_d = 1 - 0.015\,Z \qquad (7.4.2)$$

where Z is the depth of soil in metres.

7.4.2 *Dynamic simple shear testing*

Five parameters are normally measured in the dynamic simple shear test: dynamic shear stress, total vertical stress, dynamic shear strain, pore water pressure and corresponding number of cycles. From the test results one can obtain the number of stress cycles (N) for initial liquefaction, i.e. pore pressure (u) equals to the effective overburden pressure (σ_0') or for reaching a specified shear strain. The liquefaction strength is the cyclic shear stress, i.e. liquefaction strength corresponding to the number of cycles (N) to reach initial liquefaction.

Comparatively speaking, the dynamic simple shear test can simulate the process of liquefaction in natural ground more realistically than the dynamic triaxial test. However, the shortcoming of the simple shear test is that the shear stress applied to the soil sample is not uniformly distributed.

7.5 IN SITU TESTING FOR LIQUEFACTION POTENTIAL

The advantage of in situ testing lies in its ability to test soil as it exists in the field. However, due to difficulties of simulating seismic action in the field, most in situ tests for liquefaction potential measure parameters of soil behavior not directly linked to the liquefaction strength. Criteria for evaluation have been established through empirical correlation of the parameters from in situ test results with observed liquefaction events during past earthquakes.

7.5.1 *Standard Penetration Test (SPT)*

The standard penetration resistance (N) value reflects the joint effects of relative density of sand, structural characteristics of soil, lateral earth pressure at rest and overburden pressure in soil. These are pertinent factors influencing liquefaction.

The Chinese National Code of Aseismic Design for Buildings [39] specifies the following equation for identifying liquefiable sand:

Table 7.5.1. Basic N_o value of SPT.

Near/far field of earthquake	Design earthquake intensity (I) VII	VIII	IX
Near field	6	10	16
Far field	8	12	–

$$N_{cr} = N_0[0.9 + 0.1(d_s - d_w)]\sqrt{\frac{3}{p_c}} \tag{7.5.1}$$

where d_w = depth of groundwater table (m); d_s = depth of soil tested (m); p_c = fines ($D \leq 0.005$ mm) content in percent, in case $p_c < 3$, take $p_c = 3$; N_0 = an empirical basic N value that varies with the design earthquake intensity I as shown in Table 7.5.1.

The actual measured N value of soil within the depth of 15 m should be compared with the critical N_{cr} value obtained from Equation 7.5.1. The soil is considered likely to liquefy when $N < N_{cr}$.

7.5.2 *Cone Penetration Test (CPT)*

A cone penetrometer (Wang 1978, 1983) has been designed in China (Figure 7.5.1.) for assessing liquefaction potential of soils. The applicability of this penetrometer with its associated interpretation is currently supported by the Chinese experience only. The measured penetration resistance p_s is compared with the critical value p'_s calculated from Equation (7.5.2) [34]:

$$p'_s = p_{so}[1 - 0.065\,(d_w - 2)][1 - 0.05\,(d_y - 3)] \tag{7.5.2}$$

where d_w is the depth of ground water table (m); d_y is the thickness of the overlying non-liquefiable layer (m); p_{so} is the specified critical penetration resistance (MPa) for $d_w = 2$ m and $d_y = 3$ m. The value of p_{so} is shown in Table 7.5.2.

When the measured cone penetration resistance, p_s is less than the critical value p'_s, the liquefaction potential is high. Otherwise, liquefaction potential is low when p_s is greater than p'_s.

7.5.3 *Method based on shear wave velocity*

Dobry (1981) derived a relationship between shear wave velocity propagating in the soil and the threshold acceleration for a given site

$$\frac{a_t}{g} = \frac{\gamma_t\left(\frac{G}{G_{\max}}\right)}{g d_s r_d} \times V_s^2 \tag{7.5.3}$$

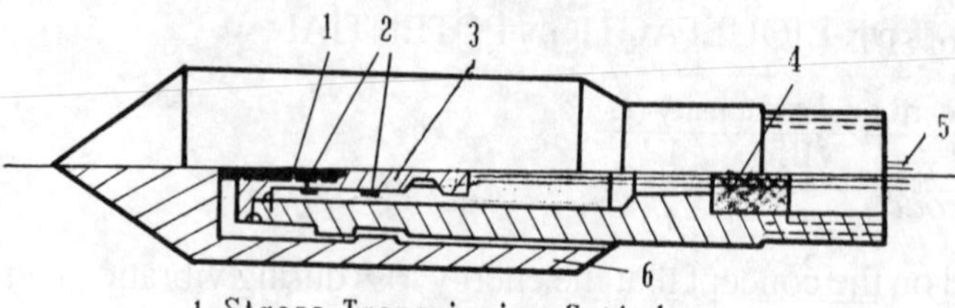

Figure 7.5.1. Standard probe of Chinese CPT.

Table 7.5.2. P_{so} versus design earthquake intensity (*I*).

I	VII	VIII	IX
P_{so} (MPa)	5-6	11.5-13	18-20

Table 7.5.3. Critical values of shear wave velocity $\overline{V}_{s,\,cri}$ versus intensity (*I*).

I	VII	VIII	IX
$\overline{V}_{s,\,cri}$ (sandy silt)	42	60	84
$\overline{V}_{s,\,cri}$ (sand)	63	89	125

where γ_t = threshold shear strain, which marks the beginning of pore pressure buildup in the soil and ranges $(1 \sim 3) \times 10^{-2}\%$; a_t = threshold acceleration defined as the peak acceleration as shear strain approaches its threshold value $\gamma = \gamma_t$; $G/G_{\max}$ = modulus reduction factor at the threshold strains γ_t; $G_{\max}$ = initial shear modulus of the granular soils measured at small strains (i.e. $\gamma \leq 10^{-4}\%$); r_d = correction coefficient for depth given by Equation (7.4.2); d_s = depth of the soil to be tested.

In light of the above relationship, a series of laboratory and in situ tests have been carried out in China, leading to a criterion in terms of V_s for assessing liquefaction potential.

For sandy silt and sand, liquefaction potential can be assessed with the following equation:

$$V_{s,\,cri} = \overline{V}_{s,\,cri}(d_s - 0.0133\,d_s^2)^{0.5} \tag{7.5.4}$$

where $\overline{V}_{s,\,cri}$ is the critical values of shear wave velocity listed in Table 7.5.3. When the measured shear wave velocity V_s is smaller than $V_{s,\,cri}$ obtained from Equation (7.5.4), the liquefaction potential is high, otherwise, it is low.

7.6 ENERGY METHOD FOR LIQUEFACTION POTENTIAL EVALUATION

7.6.1 *Deterministic approach*

This method is developed on the concept that the energy loss during vibration can be used as a parameter reflecting the dynamic response of the soil to seismic vibration. Based on the relationship between energy loss and pore pressure evaluation under dynamic loading obtained in laboratory, and on the correlation with the field data by regression analysis, the liquefaction potential can be evaluated with the following procedure (Law et al. 1990).

The increase of pore water pressure (Δu) with respect to effective overburden pressure can be expressed by

$$\frac{\Delta u}{\sigma_0'} = \alpha W_R^{\beta} \tag{7.6.1}$$

where W_R is the dissipated energy per unit soil volume in a dimensionless form, α and β are experimentally determined constants. W_R is given by:

$$W_R = F(E_I, N_1) \tag{7.6.2a}$$

and

$$E_I = \frac{\theta \times 10^{(1.5M + 4.8)}}{R^{4.3}} \tag{7.6.2b}$$

where E_I = earthquake energy arriving at the site through highly fractured bedrock; M = magnitude of earthquake; R = the hypocentral distance; N_1 = Standard Penetration Test resistance corrected to an energy ratio of 60% and an effective overburden pressure of 100 kPa.

Substituting Equation (7.6.2a) and Equation (7.6.2b) into Equation (7.6.1) yields

$$\frac{\Delta u}{\sigma_0'} = \alpha F\left[M, R, N_1\right]^{\beta} \tag{7.6.3}$$

The definition of initial liquefaction leads to $\Delta u/\sigma_0' = 1$. Consequently, Equation (7.6.3) can be split into two functions and rewritten in the following form for the condition for liquefaciton to occur:

$$\frac{T(M, R)}{\eta(N_1)} \geq 1.0 \tag{7.6.4}$$

where $T(M, R)$ is the seismic energy intensity function. For highly fractured bedrock:

$$T(M, R) = \frac{10^{1.5M}}{R^{4.3}} \tag{7.6.5}$$

and $\eta(N_1)$ is the liquefaction resistance function:

$$\eta(N_1) = 2.28 \times 10^{-10} \times N_1^{11.5} \tag{7.6.6}$$

Finally the criterion for liquefaction failure for saturated sand can be expressed as

$$\frac{10^{1.5M}}{2.28 \times 10^{-10} \times N_1^{11.5} \times R^{4.3}} \geq 1 \tag{7.6.7}$$

7.6.2 *Probabilistic approach*

For assessing liquefaction potential, a probabilistic approach based on the energy method has been developed to account for the probabilistic nature of seismic events (Law & Cao 1991). The approach is illustrated in the following using experience in Canada.

1. *Seismicity in Canada*

Based on seismictectonic setting and past recorded events, seismologists (Basham et al. 1982 and Basham & Adams, 1983) compiled seismic zoning maps for Canada. As the history of recording seismic events in Canada is relatively short, these maps should be updated from time to time with the occurrence of new events or with improved interpretation of the past events. The method proposed here treats the seismic zoning as an input data which can be changed to reflect the most acceptable practice. The illustrations in a latter section use the zoning map by Basham et al.(1982). This map (Fig. 7.6.1) has been adopted by the Canadian National Building Code since 1985. Each zone is denoted by a three-character term which is an abbreviation of the region. For each zone, the seismologists provide a recurrence relationship which expresses the probability of occurrence of an earthquake exceeding a certain magnitude. The probability of occurrence is assumed uniform throughout the same zone and is expressed as

$$N(>M) = \begin{cases} N_o \exp(-\beta M), & \text{for } M < M_u \\ 0, & \text{for } M \geq M_u \end{cases} \tag{7.6.8}$$

where $N(>M)$ = number of earthquakes of magnitude M and larger, occurring per annum in the zone; N_o, β = recurrence coefficients determined by the best of fit to the observed events using Equation (7.6.8); M_u = maximum earthquake magnitude for the zone.

Another important aspect of seismicity deals with the attenuation relationship. Work has been done by seismologists to study the attenuation relationships for intensity, acceleration and velocity. In this approach, the attenuation of the seismic energy with the hypocentral distance is used. For this, no direct information exists. Recognizing that energy attenuation is more closely related to intensity attenuation, one may therefore consider the energy attenuation coefficient (B)

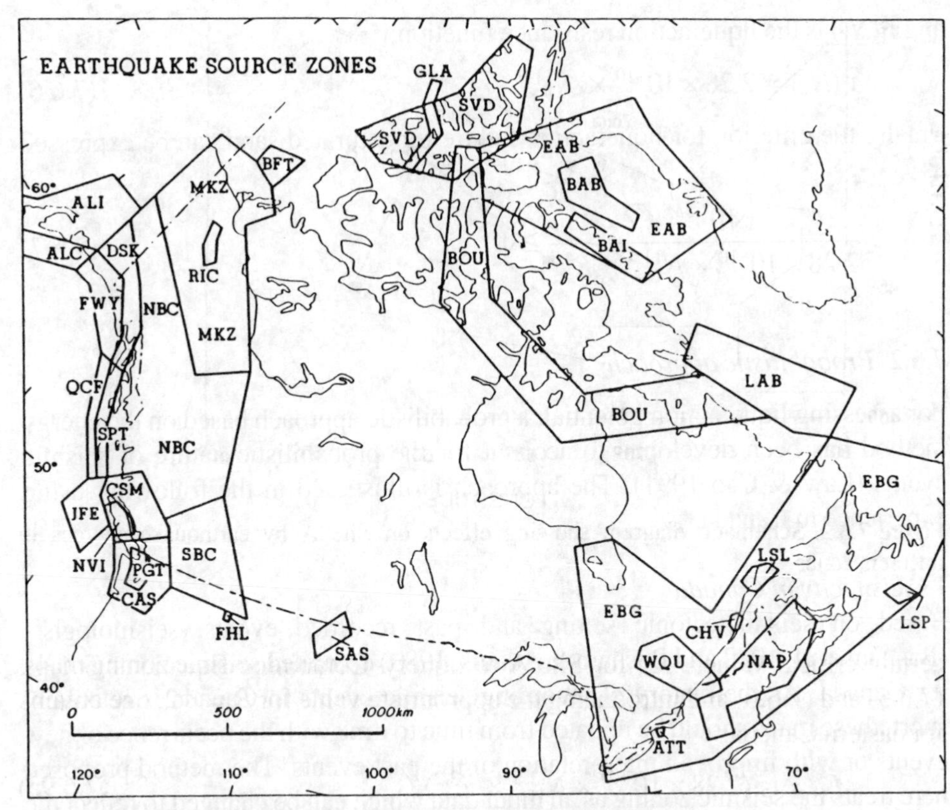

Figure 7.6.1. Seismic zoning map of Canada (Basham et al., 1982).

proposed by Hasegawa et al. (1981). For the Canadian west coast, the bedrock is of highly-fractured nature because of the interplate tectonic activities. Consequently the attenuation is high with $B = 4.3$. For eastern Canada, the rock mass is more intact and $B = 3.2$. The use of these numbers have been supported by case histories (Law et al. 1990, and Law 1991).

2. *Probabilistic assessment of liquefaction failure*

The principle adopted in this study is similar to those by Cornell (1968). The number, magnitude, and location of earthquakes influencing a given site are considered probabilistic. From the probabilistic quantities Cornell evaluated the seismic risk for a site in terms of seismic intensity, peak ground acceleration and velocity. In this approach, the probability of occurrence of seismic energy arriving at the site is evaluated. This is the energy that will cause liquefaction failure. The details are described in the following.

Consider an earthquake originating from the ith element within the jth zone (Fig. 7.6.2). The area of the element is S_i and the corresponding hypocentral distance is R_{ij} from Site A. The magnitude of the earthquake has to exceed a

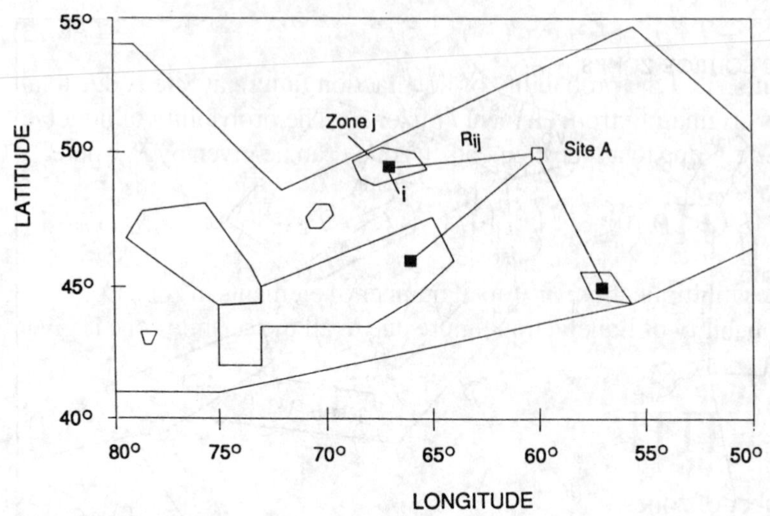

Figure 7.6.2. Schematic diagram showing effects on Site A by earthquake sources in different zones.

certain critical value, M_{cr}, for Site A to liquefy. Combining Equations (7.6.4), (7.6.5) and (7.6.6) and introducing the appropriate value for $B = 3.2$, one obtains for eastern Canada:

$$M_{cr} = \lg (2.28 \times 10^{-10} \times N_1^{11.5} \times R^{3.2})^{2/3} \tag{7.6.9a}$$

For Western Canada ($B = 4.3$):

$$M_{cr} = \lg (2.28 \times 10^{-10} \times N_1^{11.5} \times R^{4.3})^{2/3} \tag{7.6.9b}$$

From Equation (7.6.8), the annual occurrence rate of $M > M_{cr}$ is given by ν, where

$$\nu = N_o \exp(-\beta M_{cr}) \times \frac{S_i}{S_{\text{zone}}} \tag{7.6.10}$$

where S_{zone} = total area of zone j.

According to Cornell (1968), the occurrence of the major earthquakes (in this case, those causing liquefaction failures) is assumed to follow a Poisson arrival process. Then the probability for liquefaction occurring n times at Site A due to earthquakes with $M > M_{cr}$ from element i of zone j is given by $P_{ij}(n)$ where

$$P_{ij}(n) = (\nu)^n \exp(-\nu) \tag{7.6.11}$$

The probability for liquefaction not to occur, i.e. $n = 0$, is given by

$$P_{ij}(0) = \exp(-\nu) \tag{7.6.12}$$

Consequently, the probability for liquefaction to occur is given by $P_{ij}(l)$

$$P_{ij}(l) = 1 - P_{ij}(0) = 1 - \exp(-\nu) \tag{7.6.13}$$

The above results yield the probability of liquefaction failure at Site A due to all the earthquakes originating from element *i* of zone *j*. The probability of liquefaction failure at Site A due to all the elements in zone *j* can be given by P_j where

$$P_j(l) = 1 - \prod_{i=1}^{m} P_{ij}(0) \tag{7.6.14}$$

where $\prod$ denotes multiplication; m = total number of elements in zone *j*.

The total probability of liquefaction failure due to all the seismic zone is given by $P(l)$ where

$$P(l) = 1 - \prod_{j=1}^{s} \prod_{i=1}^{m} P_{ij}(0) \tag{7.6.15}$$

where s = number of zones.

In actual computation for the total probability for a given site, the integration of probability need not be carried out over all the seismic zones. From field observations (Law et al. 1990) very few earthquakes caused liquefaction for a hypocentral distance (R) exceeding 150 km. Based on calculations, the effects of earthquakes with R exceeding 200 km are practically negligible. To ensure sufficient accuracy, the calculation in this approach extends to earthquake source within 300 km from the given site.

3. *Numerical examples*

Two examples are considered here: Quebec City in Eastern Canada and Richmond in the West Coast. The entire seismic zoning map of Canada (Fig. 7.6.1) has been digitized and stored in the computer for the calculations.

With Quebec City as centre, a circle of radius 300 km is drawn as shown in Figure 7.6.3. The CHV zone is completely inside the circle while only part of the NAP and WQU zones are inside the circle. The zones inside the circle are divided up into small elements. An N_1 value is then selected for the granular soils in Quebec City. The critical earthquake magnitude, M_{cr}, that will cause liquefaction failure for the soils with N_1 is obtained from Equation (7.6.9a). The total probability of liquefaction failure, $P(l)$, from all the elements inside the circle is calculated by substituting M_{cr} in Equation (7.6.10) and carrying out the exercise from Equation (7.6.11) through Equation (7.6.15). The return period is obtained by taking the reciprocal of the total probability. The whole process is repeated for another value of N_1 and eventually, N_1 can be plotted vs the return period as shown in Figure 7.6.4. This chart is useful for evaluating the liquefaction potential of Quebec City or for microzonation purpose.

The same procedure is also applied to the case of Richmond. Because of its location, Equation (7.6.9b) is used for obtaining M_{cr}. The results are also shown on Figure 7.6.4. It is of interest to note that for the same return period, i.e. the same

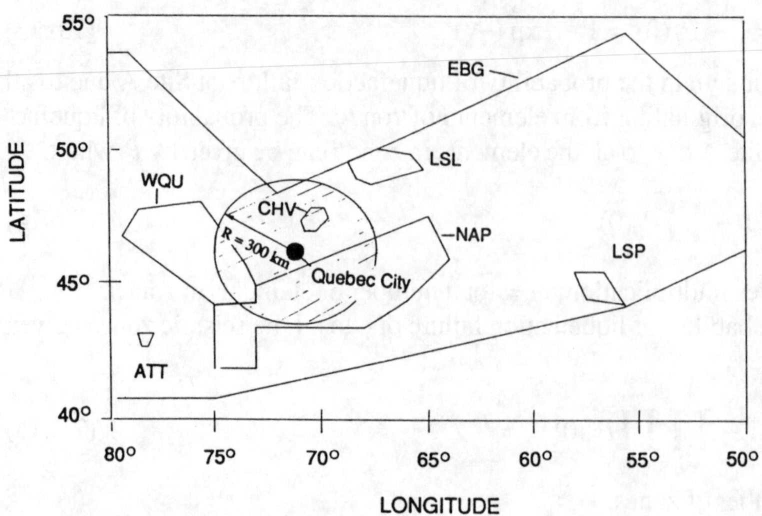

Figure 7.6.3. Defining zone of influence on Quebec City in determining the probabilistic potential.

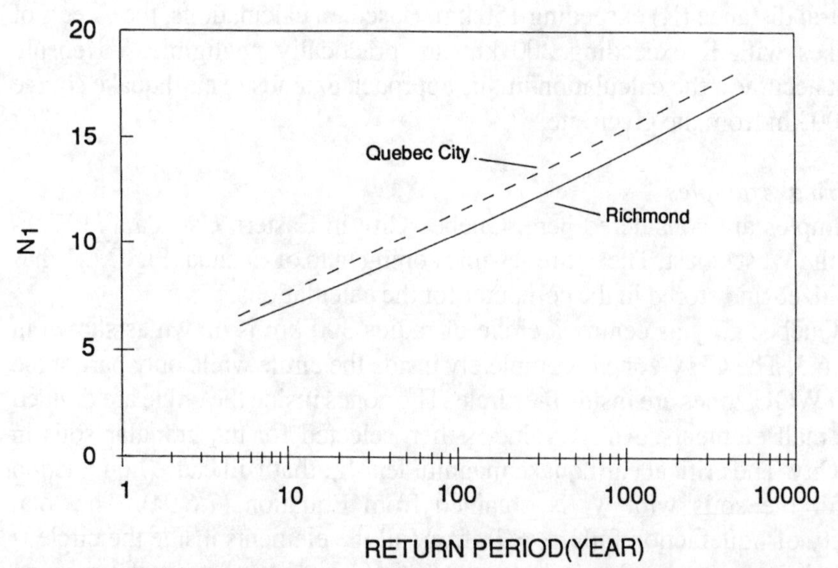

Figure 7.6.4. Relationship between N_1 and return period for Quebec City and Richmond, B.C. where N_1 is the corrected Standard Penetration Resistance below which liquefaction failure will occur.

probability of liquefaction failure, the required N_1 value is slightly lower for Richmond than for Quebec City. For the same soil characteristics, therefore, Quebec City has a higher risk of liquefaction failure than Richmond as estimated by this method. This is mainly due to the fact that energy attenuation is higher in the West Coast than in Eastern Canada.

7.7 RELIABILITY ANALYSIS FOR LIQUEFACTION POTENTIAL EVALUATION

Another probabilitic approach is briefly discussed in the following.

This appraoch is based on the stress method described in Section 7.4, i.e. seismic liquefaction is considered as a phenomenon where the seismic shear stress exceeds the dynamic shear strength of the saturated soil.

The average equivalent seismic shear stress τ_{eq} can be calculated by Seed's formula

$$\tau_{eq} = 0.65 \frac{a_{\max}}{g} \gamma h r_d \qquad (7.7.1)$$

Based on regression analysis of case records, the liquefaction strength τ_R is

$$\tau_R = 0.109\, \sigma_v'^{-0.207} N^{-0.367} d_{50}^{-0.318} \qquad (7.7.2)$$

The probability of liquefaction can be expressed by (Hardar et al. 1979)

$$P_1 = 1 - \phi \left\{ \frac{\ln\left(\dfrac{\bar{\tau}_R \sqrt{1 + V^2_{\tau_{eq}}}}{\bar{\tau}_{eq} \sqrt{1 + V^2_{\tau_R}}} \right)}{\sqrt{\ln\left[(1 + V^2_{\tau_{eq}})(1 + V^2_{\tau_R}) \right]}} \right\} \qquad (7.7.3)$$

where $\bar{\tau}_R$ and $\bar{\tau}_{eq}$ are the mean values of τ_R and τ_{eq} respectively;

$$V^2_{\tau_{eq}} = V^2_{\gamma} + V^2_{a_{\max}} + V^2_{r_d} \qquad (7.7.4)$$

$$V^2_{\tau_R} = V^2_{\sigma_0'} + V^2_N + V^2_{d_{50}} \qquad (7.7.5)$$

The definitions of the various coefficients of variation are given below, along with their typical values.

V_γ – coefficient of variation of γ, typical value is 0.03;
$V_{a_{\max}}$ – coefficient of variation of $a_{\max}$, typical range is 0.51 ~ 0.84;
V_{r_d} – coefficient of variation r_d, typical range is 0.6 ~ 2.6;
$V_{\sigma_0'}$ – coefficient of variation σ_0', typical value is 0.021;
V_N – coefficient of variation N, typical value is 0.30;
$V_{d_{50}}$ – coefficient of variation d_{50}, typical value is 0.12.

The probability of liquefaction for a given site can be obtained from Equation (7.7.3) after the values of τ_{eq}, τ_R and the six coefficients of variation are evaluated.

7.8 PREDICTION OF LIQUEFACTION HAZARD

Whether or not a site is liquefiable is not the complete answer to earthquake

Table 7.8.1. Severity of liquefaction.

Degree of liquefaction D_L	<4	4-10	>10
Severity of liquefaction	Light, occasional sand boils, insignificant settlement	Many sand boils, obvious ground settlement, structural damages occur	Hazardous, large area of sand boils, serious structural damages

resistant consideration. Different liquefiable sites may result in different severity of liquefaction hazards. The complete answer is how severe the liquefaction hazard would be, in addition to the preliminary answer 'yes' or 'no'.

The severity of liquefaction hazard is subject to many factors which can generally be represented by the coefficient of liquefaction resistance F_{LR} defined as

$$F_{LR} = \frac{\tau_R}{S_r} \tag{7.8.1}$$

where τ_R is the liquefaction strength as determined by Equation (7.7.2) S_r is the ratio of the peak seismic shear stress to the effective overburden pressure σ_v' as given by:

$$S_r = \frac{a_{\max}}{g} \times \frac{\sigma_V}{\sigma_V'} \times r_d \tag{7.8.2}$$

where r_d is again the factor to account for depth as given by Equation (7.4.2).

The degree of liquefaction D_L can be expressed (Iwasaki, 1986) as:

$$D_L = \int_0^{15} (1 - F_{LR}) W(Z) dZ \tag{7.8.3}$$

where $W(Z)$ is the weight function of the depth (Z) of soil strata given by:

$$W(Z) = 10 - \frac{2}{3} Z \tag{7.8.4}$$

and Z is the depth in metres.

Table 7.8.1 shows the relationship between the degree of liquefaction D_L and the severity of liquefaction hazard [41].

7.9 DUAL EFFECT OF LIQUEFACTION ON EARTHQUAKE DAMAGE

7.9.1 *Evidence from historical events*

It has been shown by numerous investigations on damage of buildings on the ground caused by strong earthquakes that among sites of the same epicentral distance and topography, those with liquefaction suffered lighter damage than the others without liquefaction. In general, most liquefied area are of intensity VI ~ IX. Very few of them can be seen in area of intensity *X*, and almost none of them fall in the category of I = XI. Observations in fact show that liquefaction may decrease earthquake damages and hence lower intensity even if the magnitude is high. A typical example is Niigata earthquake (1964) of $M = 7.4 \sim 7.5$. Due to liquefaction occurrence over a vast area, the intensity was only about VIII.

On the other hand, when the earthquake magnitude is small, liquefaction related damage becomes the dominant component. A site suffering from liquefaction failure will, therefore, display a higher intensity than the one without liquefaction failure.

7.9.2 *Conceptual explanation*

Liquefaction is a failure of ground soil which may cause ground settlement, landslide and subsidence. However, as soon as it occurs, some new phenomena may be produced, which may reduce the capacity of ground motion. These phenomena are described in the following:

1. The liquefied layer acts as a liquid medium which can isolate seismic shear waves transmitted from the solid bedrock.

2. The seismic energy trasmitted to the liquefied layer will be immediately absorbed and exhausted by way of producing sand boils or landslides. The residual seismic energy is largely weakened and its power of further distruction is reduced.

3. From any dynamic tests of soil in the laboratory (say dynamic triaxial), it can be easily noted that before liquefaction occurs, the soil sample can take a significant dynamic stress. As soon as liquefaction occurs, the soil is softened and can no longer take on and transmit any significant stress. This reduction of stress level being transmitted by the liquefied soil mass helps to lower damage potential.

As a general observation, soil liquefaction is a hazard to the stability of sites, ground facilities and structures. In certain disastrous earthquake of high magnitude, however, liquefaction may in fact save lives because it reduces damage potential and causes less buildings to collapse.

CHAPTER 8

Landslides and slope stability under seismic action

The purpose of this chapter is to provide guidelines of dealing with inclined terrains or slopes during siting in earthquake zones. The main concern is landslides and slope stability under seismic action.

Slopes that require seismic consideration in stability assessment include the following:

1. Natural slopes in cohesionless or low cohesion soils with a slope angle greater than the angle of repose under the submerged condition.
2. Clay slopes with sand or silt seams inclined in the same direction of the slope.
3. Slopes of rock outcrop which has a dip angle less than the slope angle and in the same direction as the slope.
4. Slope of soils overlying bedrock of a retrograde nature and unfavourable inclination.

For all the above cases stability should be evaluated by means of the pseudo-static limit equilibriumm method or by dynamic stress-strain analysis. For cases 1, 2 and 4 additional evaluation should be conducted on the likelihood of liquefaction failure in the granular soil mass, granular soil seams and along the interface between the soil and the bedrock.

8.1 EARTHQUAKE INDUCED LANDSLIDES

Under seismic action, an inclined soil/rock mass may be triggered to slide if it is already approaching a certain critical state under the static loading conditions. This type of landslide is thus defined as earthquake induced landslide. The inclined soil/rock mass may be a natural deposit or man-made earth structures such as open-cut, embankment, dike or dam, etc. However, for siting in earthquake zones we are mainly concerned with earthquake induced landslides in natural deposit. Most of the landslides induced by earthquakes took place in seismic zones of intensity (*I*) equal to or greater than VIII (MMS). Very few landslides have been caused by earthquakes in zones of intensity VII or less (I ≤ VII MMS).

8.2 GENESIS AND CATEGORIES OF EARTHQUAKE INDUCED LANDSLIDES

Based on siting requirements in earthquake zones, earthquake induced landslides may be grouped into two categories.

8.2.1 *Earthquake induced landslides in plain areas*

Landslides induced by earthquakes do not necessarily occur only on slopes. In plain areas, landslides may occur at certain sloped ground such as sea beach, river bank, undermined ground, back-filled abandoned valley, etc. However, earthquake induced landslides in plain areas are often related to soil liquefaction. When an inclined liquefiable layer exists as an underlying stratum, the overlying soil may slide even though the ground surface is level. The methodology of assessing liquefaction potential is given in Chapter 7.

The internal factors causing landslides in plain area are the geometry of the sloped ground and soft and weakened interfaces due to groundwater intrusion. The external factor is mainly the amplified ground shaking. Shaking on horizontal ground is likely the result of repeated superposition of vibration amplitudes of different phases including those directly from the seismic source and from structures set into motion during the earthquake. If such action takes place at the sudden change of topography or stratigraphy, intensified and irregular ground shaking may occur and cause long range waving deformation of sliding ground.

8.2.2 *Earthquake induced landslides in mountainous areas*

The common seismic factors causing landslides in mountainous areas are as follows:

1. Horizontal seismic acceleration acting on the slope may increase sliding force of soil mass, or cause additional pore water pressure in the soil, which reduces the resisting force to sliding.
2. Surface faulting, if any, that goes across a slope may form an open crack extending to underlying weak layers. This facilitates infiltration by groundwater or precipitation into those weak layers and results in sliding.
3. The amplification of ground motion by the overlying deposit causes differential movements between the overlying loose or soft soil and the solid bedrock. Such movements may result in relative motion on the sliding interface.

8.3 STABILITY EVALUATION OF EARTHQUAKE INDUCED LANDSLIDE

For siting in earthquake zones, the stability problem under the direct action of earthquake can be evaluated both qualitatively and quantitatively.

8.3.1 *Qualitative evaluation*

The basic principle of a qualitative evaluation is to make preliminary assessment and prediction on slope stability according to the natural circumstances of slope, the existing conditions of soil and rock, and the ground water regime. Three grades of stability are suggested as shown in Table 8.3.1 in which factors A, B and E are used for plain areas, whereas factors C, D and E are used for mountainous areas.

8.3.2 *Quantitative evaluation*

Quantitative evaluation is necessary for siting of large and important projects. Conclusions can be drawn by synthesizing both qualitative and quantitative evaluations. The process is shown in principle in Figure 8.3.1. Siting for an ordinary projects, however, does not necessarily follow the whole procedure shown in the figure The engineer may make his/her own judgement to start with either the quantitative or the qualitative method. In case the results from the qualitative and the quantitative evaluations are inconsistent, further analysis (e.g. the reliability analysis by probabilistic method, etc.) is required to reach a comprehensive assessment.

1. *Pseudo-static analysis*

This analysis is an ordinary method for analyzing slope stability under seismic action. The major concept is as follows: Take the inertial seismic action (which varies with direction and time) to be a time-independent constant force (i.e. pseudo static) acting on the soil/rock mass. Such a pseudo-static force is considered to be the product of seismic coefficient K_h and the weight W of the soil mass

Table 8.3.1. Indices and grades for slope stability evaluation.

Factors indices grades	Slope and lithology Slope A	Soils B	Slope C	rocks D	Depth of perched water (m)
Unstable or critical	>25°	Loose sand silty clay, hydraulic fill	>50°	Fractured fault zone, strongly weathered rock with soft clay filler	<3
Less stable or understable	10-25°	Lose cohesive soil, newly deposited earth	30°-50°	Moderately weathered rock with fissures of 20-50 cm spacing with little mineral fillers	3-10
Relatively stable	5°-10°	Coarse sand, gravelly soil, compacted cohesive soil, and deposited earth	20°-30°	intact and solid rock, slightly weathered fewer joints and fissures	>10

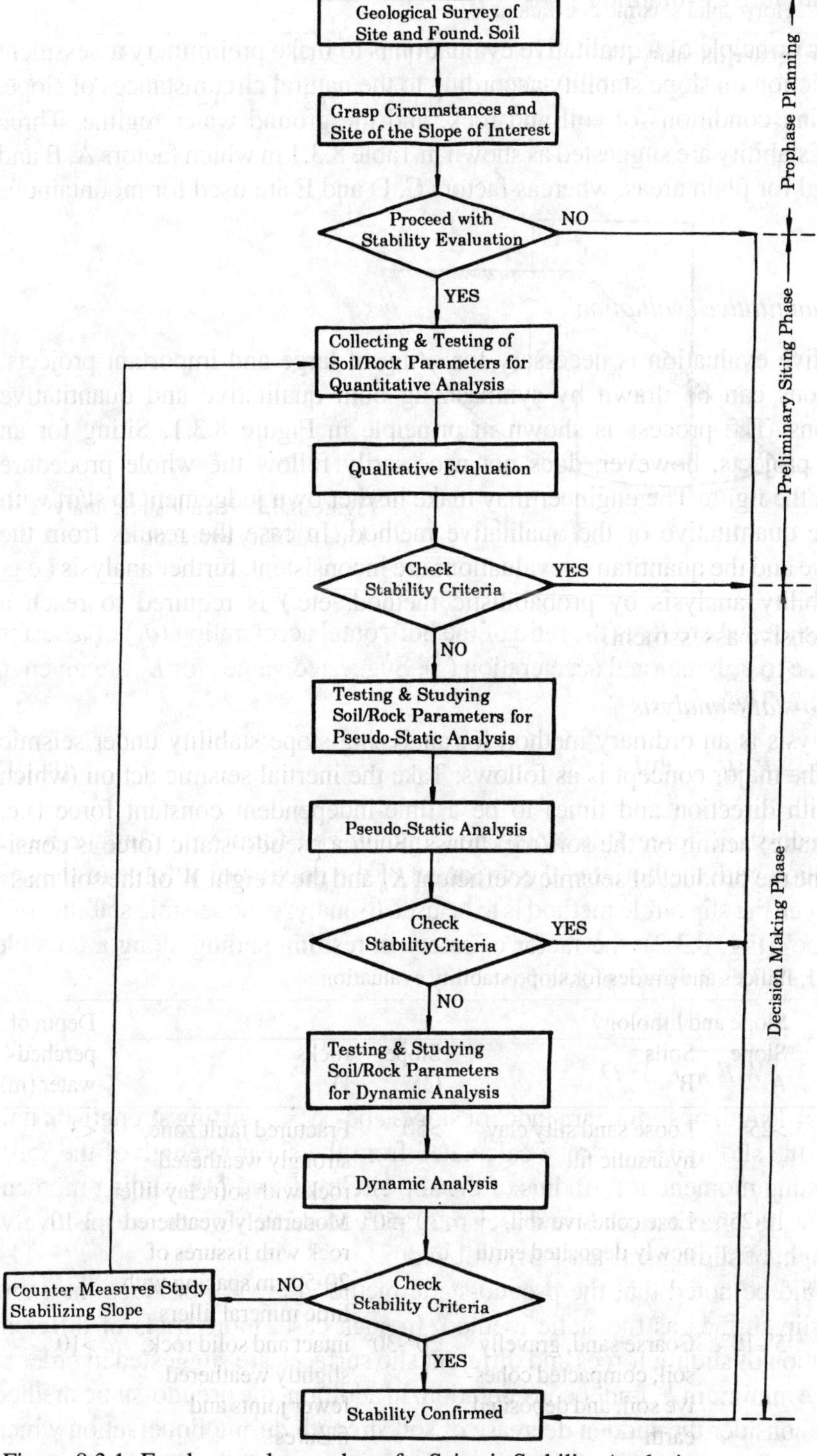

Figure 8.3.1. Fundamental procedures for Seismic Stability Analysis.

Table 8.3.2. Horizontal seismic coefficient K_h.

Design earthquake intensity (I)	VII	VIII	IX
K_h	0.1	0.2	0.4

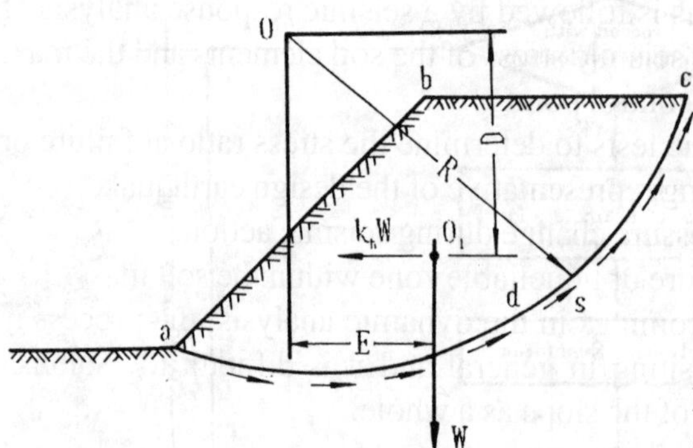

Figure 8.3.2. Pseudostatic analysis of slope stability evaluation.

likely to slide, where K_h is the ratio of the horizontal acceleration (a_h) of a design earthquake to gravitational acceleration (g). Suggested values for K_h are given in Table 8.3.2. Thus,

$$P = K_h W = \frac{a_h}{g} W \tag{8.3.1}$$

The pseudo-static force P is acting horizontally and forming an unfavourable combination with the horizontal component of the gravitational force of the soil mass. When the slip circle method is to be used to analyze the seismic stability of a slope 'abc' (Fig. 8.3.2), the factor of safety in resiting sliding along a possible sliding arc 'adc' can be expressed as:

$$F_s = \frac{LSR}{WE + K_h WD} \tag{8.3.2}$$

where L = length of sliding arc 'adc' of slope 'abc'; S = resisting strength on unit width of the sliding mass (to be calculated from the shear strength of the soil); R = resisting moment arm (radius of the slip circle); E and D = sliding moment arms due to gravitational and seismic pseudo-static forces, respectively; W = weight of sliding soil mass per unit width.

It should be noted that the pseudo-static method may not produce the same critical slip surface as the static method. In such case, more trials of different combination of sliding forces and different slip surfaces are suggested in order to obtain the minimum F_s under consideration. In addition, the pseudo-static method does not consider the sudden decrease of soil strength during liquefaction which may cause sliding of a liquefiable slope. For this type of slope, the pseudo-static method will yield results on the unsafe side. In this case, the dynamic analysis is suggested for further assessment.

2. *Dynamic analysis*
The following procedures are generally suggested for a dynamic analysis.

1. Carry out a static analysis of the slope system to determine the static stress components of soil elements. This is followed by a seismic response analysis of the slope system to determine the seismic stress of the soil elements and the max. horizontal acceleration at certain control points.
2. Perform appropriate dynamic tests to determine the stress ratio at failure or at liquefaction under cyclic loading representative of the design earthquake.
3. Calculate the pore water pressure change during seismic action.
4. Determine, if any, shear failure or liquefiable zone within the soil mass.

There are inevitably some shortcomings in the dynamic analysis: the process is too complicated to be used for siting in general and it is difficult to establish criteria for assessing the stability of the slope as a whole.

8.4 GEOTECHNICAL INVESTIGATION FOR SLOPE STABILITY EVALUATION

A geotechnical investigation contains two major parts:

1. The engineering geological environment and testing of shear resistance parameters of the sloped ground.
2. The ground water regime which is likely to change the slide-resisting capability of the slope.

Investigation may be started with the collection of engineering geological maps and soil profiles of the major section of the slope under investigation. Aerial photos may help identify existing landslides. If landslides induced by historical earthquakes exist, study the surrounding topography, geomorphology and soil/rock parameters to provide supporting references to the seismic stability evaluation of existing slopes.

For investigation of the sloped or hilly areas in earthquake zones, emphasis should be put on the evidences or traces of historical events, if any, such as surface faulting, ruptures, waving deformation of ground surface and sand boils/sand seams, etc. A particular conclusion should be drawn as to whether any of these traces caused any past sliding of the sloped ground. Field work such as borehole drilling and sampling, digging trenches and pits, sounding and penetration tests, geophysical surveys and ground water monitoring are useful towards this end. In addition, stratigraphic characteristics such as fractured zones, fissures and joints, especially the weak and soft sandwich layer in rock formations should be investigated in details. If necessary use downhole camera or television for the investigation.

Table 8.4.1. Suggested shear strength test methods.

Current state of the slope	Soil/rock character in potentially sliding zone	Suggested shear strength test method
Active slope currently undergoing creeping	Clayey or residual soils	Repeated direct shear test to determine the residual strength or dynamic test at stress level corresponding to the acceleration of the design earthquake
	Liquid-plastic-muddy clay	Unconsolidated undrained test
	Plastic solid soil or gravelly interface between rock formations	Multiple shear test with folded shearing surface
Natural slope	Non-homogeneous or stratified soils	Large scale in situ shearbox
	Saturated clayey soil	Unconsolidated undrained shear test or consolidated undrained test and long term strength test for deeply embedded clayed soil
	Saturated, medium dense or looser soils with low to medium cohesion	Unconsolidated undrained shear test and liquefaction test

Laboratory testing is normally focused on determing the shear strength of soil/rock of the slope. Strength parameters C and ϕ can be obtained by using methods appropriate to the conditions of the slope and its material as suggested in Table 8.4.1.

The groundwater regime in and surrounding the slope should be investigated with emphasis on shallow embedded water especially for the perched water in the sloped ground and locally pressurized (artesian) water. The direction and rate of flow of streams, swamps, and springs should also be noted.

For siting important and large projects in slope areas or under-stable ground (Table 8.3.1), further investigation on run-off, erosion by surface flow and groundwater monitoring are necessary. More detailed study on the interaction of groundwater and sloped soil/rock mass is sometimes needed to solve the following problems:

1. Whether the drainage condition of soil/rock slope is sufficient.
2. Whether the groundwater regime does any harm to slope stability (such as mechanical erosion, chemical leaching etc.).
3. Whether excess pore water pressure exists (which may decrease effective stress in soil and cause failure).
4. Whether there is any intrusion of surface water into the soil/rock through ground fissures, shrinkage cracks, trenches, culvert or any other man-made excavations.

If any of these exists, take soil samples during the unfavourable season, and measure the shear strength of the soil and evaluate pore water pressure influence.

CHAPTER 9

Ground waving and its damaging effect

9.1 CASE HISTORIES

Numerous investigations on the damaged areas of strong earthquakes have presented many unusual patterns of destruction which are incredible but have actually happened. Among them, ground waving is the most significant one in the sense of seismic hazard and its mitigation. The authors prefer using this the term 'ground waving' because it properly describes the real behaviour of ground movement during strong earthquakes, which is portrayed by the wavy form of permanent deformation of the ground surface after the earthquake. Eyewitnesses in the field during some strong earthquakes, such as the Xingtai earthquake (M = 7.2, March 22, 1966, North China) which took place during daytime, reported seeing noticeable waving of ground just like the patterns left permanently in the field.

The mechanism and geometric shapes of the structural damages caused by ground waving are so strange that they appear to be distinctively different from those in the current methodology of seismic design. Conventionally, seismic force acting on a structure is visualized as the product of the mass of the structure multiplied by the seismic acceleration from the design response spectrum. However, this is not applicable to some of the ground waving damages as shown in the following examples.

9.1.1 *Torsional ground waving*

An illustrative example of ground waving damage occured in Tangshan earthquake (M = 7.8, 1976) is shown in Figure 9.1.1(A). The rails in this plate were damaged in a waving manner by overwhelmingly kinematic motion of ground surface which exhibits very clear surface wave form – that of horizontally rotational Love wave. Although there were no strong seismic arrays installed in this area and even the available strong seismographs were all out of scale, the permanent ground movements did retain the actual behaviour of the ground surface at that particular locality.

Figure 9.1.1. (A) Rails distorted vertically due to Rayleigh surface wave motion, (B) Train and rails distorted horizontally due to Love surface wave motion.

Figure 9.1.1(B) shows another pattern of ground waving – vertical rocking due to surface Rayleigh wave which occurred at different places of different site conditions during the same event.

These damages clearly demonstrate the fact that the rails were distorted not directly by seismic force acting on them, because their mass is too small to exert destructive seismic inertial force. It is obvious that the rails were attached to the sleepers which were mainly embedded in the ballast of the roadbed. Therefore, they were actually attached to the ground and thus deformed completely together with ground waving. In situ measurement on surface wave parameters show that the distorted rails appeared to be shorter in wave length than that of Love wave or Rayleigh wave. This may be due to the coupled oscillation with a compound rigidity of both the rail and the ground soil.

9.1.2 *Stationary wave destruction*

It was found through site investigations that when seismic wave travels along the surface of a confined area like river channel, small lake or basin deposit, reflected secondary waves will be induced and propagate back and forth within the limited boundaries. Eventually, as the progressing wave coincides with the reflected wave in the same phase, amplified magnitude of oscillation will take place at all the peaks and will persist for a certain time duration. This phenomenon can be demonstrated even with a simple simulation in the laboratory.

Figure 9.1.2(A) is one of the highway bridges in Tangshan area totally damaged in the manner of throwing over and overlapping together of the reinforced concrete plate girders in the longitudinal direction without any transverse (river stream direction) displacement at all. This pattern of damage is very representative of the river bridges of small and medium size destroyed in high intensity zones.

A striking contrast can be seen in Figure 9.1.2(B) where a group of single storey long span masonry warehouses just behind the wholly collapsed small bridge, stood still without major damage. The possibility of ground resonance damage was ruled out because there was not much difference in natural frequency between the bridge and the house. However, their behaviours were so different.

Theoretically, stationary wave may take place when the group velocity of a surface seismic wave packet coincides with the velocity of the seismic wave in the overburden layer. Taking the conditional equation of Love wave and letting v_1, v_2 represent the horizontal displacements above and below the interface along Y-axis, we have

$$v_2\sqrt{1-\frac{V_L^2}{V_{s2}^2}} = v_1\left(\sqrt{\frac{V_L^2}{V_{s1}^2}-1}\right)\tan\left(\frac{\omega h}{V_L}\sqrt{\frac{V_L^2}{V_{s1}^2}-1}\right) \tag{9.1.1}$$

where V_{s1} and V_{s2} are shear wave velocities of the overburden and the half-space, respectively. V_L is the velocity of Love wave.

Obviously, solution to Equation (9.1.1) exists only when $V_{s1} < V_L < V_{s2}$. Due to dispersion of seismic wave, V_L varies with frequency or wave length and becomes so small that V_L approaches V_{s1}, resulting in a magnification of the stationary Love wave.

Based on the research on *SH* and Love waves in a homogeneous, isotropic, linear and elastic layered medium, Dravinski & Trifunac (1979) summarized that the contribution of local stationary waves to the total energy density spectrum is significant within the thickness of the top layer.

Figure 9.1.2. (A) Highway bridge overthrown by locally intensified stationary wave motion within the river channel only – whereas the large-span masonry warehouses on the background stood still, (B) Lan-He bridge of 36 spans totally fell down in longitudinal direction.

9.1.3 *Rhythmic destruction*

Rhythmic destruction is a term to describe the type of earthquake damages that selectively occur in structures spaced at a regular distance.

Figure 9.1.3(A) is an actual example of rhythmic destruction in Tangshan earthquake where six masonry apartment buildings with basically equal distance

Figure 9.1.3. (A) Rhythmic destruction: Six residential buildings of Tangshan Cement Company suffered from earthquake quite differently. Among them, (B), (D), (F) totally vanished, whereas (A), (C) were safer and (E) partially damaged.

Figure 9.1.3. (B) Six single storey houses of the same design and construction damaged intermittently with a spacing roughly equal to surface Rayleigh wave length – B, C, E, F totally collapsed; but A, D were safe.

apart were constructed at the same time, with the same building materials and by the same construction team. However, they were damaged in striking contrast – three of them totally collapsed and the other three stood still with partial damage. Figure 9.1.3(B) shows a similar example of the same phenomenon in which six identical houses were built at equal spacing. What is more evident here is two single storey houses were intact without any damage while the other four, located alternately between the intact ones, totally vanished.

Rhythmic destruction can be simply understood by considering the motion of surface wave packet composed of stationary waves. When local amplification occurs, such as caused by stationary wave action, the peaks are magnified significantly while the webs remain at zero amplitude. In the case of strong vertical motion, the peaks of upward half cycles opposite to gravitational acceleration will cause much more damage than the downward half ones. Any structure or building on the ground situated at the points where magnified peaks occur, will undergo severe translational shaking, rocking or rotational waving.

Rhythmic destruction may also be resulted from the reflection of surface wave within a partially confined curvilinear barrier such as a dike along a convex river bank, etc. Travelling surface waves reflected from those boundaries are likely to be concentrated along the central area or to coincide in equal phase to form a standing amplified wave action. Figure 9.1.4 is an example among many showing a half-looped site confined by the convex bank of a branch of Lanhe river. The white stripes were actually composed of numerous spots of sand boils due to soil liquefaction under intensified surface wave motion. Spacings between the white stripes of sand boils, are clearly equal to the spacing between the centre lines of the collapsed single storey houses B-C and E-F as shown in Figure 9.1.3(B). In fact, the site in Figure 9.1.3(B) is actually situated at the top-left river bend shown in Figure 9.1.4 which was taken from the air. The picture shows that the white stripes of sand boils and the centre lines of the collapsed houses fall on the same alignment, even though houses (A) and (B) are not as clear from the aerial view due to the shading of trees.

Figure 9.1.4. These sites within the river-bends experienced much more severe damages than other places outside the river-bends. White stripes are numerous spots of sandboils on the ground taken by aerial photogrammetry showing locally intensified ground waving.

Figure 9.1.5. An example illustrating incompatibility between lengthy building and ground waving.

9.1.4 *Influence of ground waving on lengthy structures*

During strong earthquakes, a structure and the ground on which it is situated will follow the waving motion together as a single unit. Incompatibility in motion behaviour between the structure and the ground will occur if the total length of structure is much larger than the surface wave length of the ground. Linear or lengthy structures are, therefore, vulnerable to strong waving of the ground surface. Linearly shaped buildings longer than the surface wave length of the ground may experience striking by the ground at its end or distortion at its expansion joints. Figure 9.1.5 shows the tilting of the end segment of an industrial building struck by ground waving at its lower part close to the ground level.

9.2 BRIEF REVIEW OF RELEVANT RESEARCH AND KNOWLEDGE

9.2.1 *Torsional motions*

In many countries, seismic design codes or seismic regulations for buildings such as ATC 3-06 of USA and NBCC-90 of Canada etc, have specified a definite requirement of checking seismic resistance with torsional moment. However, the torsion specified in those existing provisions is only defined as that resulted from the uneven mass distribution and the eccentric displacement of the mass of the building. In fact, it has been widely recognized that in earthquake damaged areas many axial-symmetric masonry structures like chimney stalks, cylindrical masonry wall supporting water tanks etc, irrespective of their size, did appear to be horizontally sheared into segments due to horizontally torsional movement of the ground. In addition, in many other cases of torsional damage, the centre of rotation is outside the structure. Therefore this kind of damage is obviously outside the scope of the existing seismic provisions.

In recent years, more and more attention has been paid to the torsional components of seismic waves by many researchers.

Niazi (1982) first obtains an inferred rotational motion using decomposed travelling waves from a 3-*D* accelerogram of a strong motion array in EL Centro during the Imperial Valley earthquake in 1979.

Trifunac (1982) carries out systematic research on both translational and rotational components of seismic ground motion by inducing body wave into the elastic half space. Lee & Trifunac (1985) further develop synthetic torsional accelerogram from translational data.

Oliveira & Bolt (1989) estimate rotational components of seismic waves by using densified strong motion array SMART-1 in Taiwan and arrive at the conclusion that when Curl $u > 0.0001$ rad, significant effects of rotational components generally appear in engineered structures.

Wang et al. (1983) [35] conceptually demonstrates that whenever dipping interface exists between the overburden and the underlying layer and that shear wave velocity of the former is smaller than that of the latter, rotational motion of the ground surface must occur.

Recent advances in this aspect have been made by Boffi & Castellani (1991), Jin & Liao (1991). The latter develop synthetic design rotational and rocking components with synthetic translational Fourier spectra for a known source with given magnitude and epicentral distance.

Theoretical approach generally considers travelling wave in the elastic homogeneous layered half-space in which the rotational displacement vector at the nth layer is:

$$\Omega_n(x_1, x_2, t) = \frac{1}{2}\left(\frac{\partial u_{3n}}{\partial x_2}\bar{x}_1 - \frac{\partial u_{3n}}{\partial x_1}\bar{x}_2\right) \tag{9.2.1}$$

or

$$\Omega_n(x_1, x_2, t) = \Omega_{1n}(x_1, x_2, t)\bar{x}_1 + \Omega_{2n}(x_1, x_2, t)\bar{x}_2 \tag{9.2.2}$$

where $\bar{x}_1$, $\bar{x}_2$ are the unit vectors along x_1 (horizontal), x_2 (vertical) axes respectively. u_{3n} is the displacement of a point in the nth layer along the x_3 axis (horizontal and normal to the paper) and

$$\Omega_{1n} = \frac{1}{2\rho_n V_{sn}^2}\tau_{2,3n} \tag{9.2.3}$$

$$\Omega_{2n} = \frac{i\omega}{2c(\omega)} u_{3n} \tag{9.2.4}$$

where ρ_n is the mass density. V_{sn} is the shear wave velocity of the nth layer. $\tau_{2,3n}$ is the shear stress on x_2, x_3 axial plane. $c(\omega)$ is the phase velocity along horizontal axis x_1.

Equations (9.2.3) and (9.2.4) demonstrate that when shear wave polarizes through an interface, it will result in two torsional motions: one is the torsion around horizontal axis which is directly proportional to the shear stress; the other is a rotation which is subject to the out-of-plane displacement. Since there is no stress on the free ground surface, Equation (9.2.4) is the only expression for rotational motion on the ground surface.

Therefore, for Love wave, the rotational acceleration in terms of Fourier spectrum $\phi_2(\omega)$ can be obtained from the out-of-plane acceleration Fourier spectrum $a_3(\omega)$, i.e.

$$\phi_2(\omega) = \frac{i\omega}{2c(\omega)} a_3(\omega) \tag{9.2.5}$$

For Rayleigh wave on ground surface induced by P-SV waves, two translational and three torsional components in the nth layer of the elastic half space are:

$$u_{1n} = \frac{\partial\phi_n}{\partial x_1} + \frac{\partial\psi_n}{\partial x_2}$$

$$u_{2n} = \frac{\partial\phi_n}{\partial x_2} - \frac{\partial\psi_n}{\partial x_1}$$

$$\Omega_{1n} = 0$$

$$\Omega_{2n} = 0$$

$$\Omega_{3n} = -\frac{1}{2}\nabla^2\psi_n \tag{9.2.6}$$

where ϕ_n, ψ_n are scalar potential function and vector potential function of the nth layer, respectively. ψ_n satisfies the standard wave equation, thus

$$\Omega_{3n} = \frac{\omega^2}{2V_{sn}^2}\psi_n \tag{9.2.7}$$

Similarly, the Fourier spectrum $\phi_3(\omega)$ of torsional (rocking) acceleration around the x_3 axis on the ground surface can be obtained from the Fourier spectrum $a_2(\omega)$ of translational acceleration along the x_2 axis:

$$\phi_3(\omega) = \frac{\omega}{ic(\omega)} a_2(\omega) \tag{9.2.8}$$

Thus by inverse transformation of Fourier spectra in Equations (9.2.5) and (9.2.8), the time histories of both rotational acceleration $\phi_2(t)$ and rocking acceleration $\phi_3(t)$ can be obtained.

The above procedures are summarized by the following steps:

1. Produce out-of-plane Fourier spectrum $a_3(\omega)$ and vertical Fourier spectrum $a_2(\omega)$ by means of the semi-emperical method for simulating design accelerograms (Liao et al. 1989).

2. Calculate the approximate horizontal apparent wave velocity or phase velocity $c(f)$ which can be evaluated through the relationship between the group velocity $U(f)$ and equivalent phase velocity $c_e(f)$:

$$c_e(f) = \frac{2\pi f}{K(f)} \qquad (9.2.9)$$

where $K(f)$ is the equivalent wave number

$$K(f) = \frac{2\pi f_{max}}{V_p} = 2\pi \int_{f_{max}}^{f} \frac{df}{U(f)} \qquad (9.2.10)$$

It is assumed here that the phase velocity of the highest frequency (say $f_{max} = 25\ Hz$) is the P-wave velocity (V_p), and the group velocity for P-wave $U_p(f) = 6$ km/s. Further, $c_e(f)$ correlates with the horizontal apparent wave velocity $c(f)$ as follows:

$$c(f) = \frac{c_e(f)}{\sin\theta} \qquad (9.2.11)$$

where θ is the incident angle of seismic wave which has different values, say 30°, 60°, 90°, dependent on the epicentral distance Δ. If Δ is large, $\theta \simeq 90°$, then

$$c(f) \simeq c_e(f) \qquad (9.2.12)$$

3. Based on Equations (9.2.5) and (9.2.8) and the value of $c(\omega)$, the Fourier spectra of horizontal rotation $\phi_2(\omega)$ and of vertical rocking $\phi_3(\omega)$ can be obtained by substituting the values of $a_3(\omega)$, $a_2(\omega)$, respectively.

4. The required time histories of both rotational acceleration $\phi_2(t)$ and vertically torsional (rocking) acceleration $\phi_3(t)$ can be obtained through inverse transformation of Fourier spectra $\phi_2(\omega)$ and $\phi_3(\omega)$.

However, the above treatment should be considered as a theory, because the torsional components thus separated from the synthetic translational motion are too small to cause overwhelming damages as shown in Figure 9.1.1. There was no direct measurement to prove the intensiveness of the torsional movements. In addition, there are some difficulties to apply the above theoretical approaches:

1. Strong motion array is very rare and can hardly be installed to meet practical need, let alone enough records are to be accumulated.

2. The use of synthetic rotational and rocking components seems to be realistic

only when there are enough strong motion records to provide statistical data on amplitude spectrum and phase spectrum. At present, the profession does not have enough of such records. The actual situation is even worse when one realizes that the source parameters M (magnitude), L (fault length), h (focal depth) etc. are difficult to predict. Furthermore, the angle of incident wave which affects the horizontal apparent wave velocity is also difficult to estimate in advance.

3. Theoretically speaking, both rotational and rocking motion can be induced anywhere through the polarization of seismic shear wave on the top layer whose shear wave velocity is smaller than that of the underlying stratum. However, observations show that disastrous and irresistible damages due to rotational and rocking motion occurred only at some highly specific sites where apparently torsional movements were locally magnified.

4. Many additional factors may induce local magnification of torsional movements, e.g. dipping layers, sedimentary basin and topographic irregularities causing diffraction and scattering of surface waves other than normally induced torsional components. All these may significantly influence locally intensified motion (Dravinski 1984).

9.2.2 *Stationary wave destruction*

The mechanism of stationary wave has been theoretically developed. However, seismic destruction cases caused by surface stationary waves are extremely specific in terms of local topographic and geological features of a certain site and can hardly be analytically formulated or numerically modelled for prediction. In fact, the propagation of seismic waves on ground surface in many cases is a complicated combination and superimposed action of diffracted, scattered and reflected waves in layered strata. From those wave packets a stationary motion may eventually be formed. Therefore, this kind of seismic hazard can only be understood from historical earthquake on a particular site so that mitigation of future hazard on the same or similar site can be made.

9.2.3 *Rhythmic destruction*

This type of earthquake damage has not been widely recognized so far in earthquake geotechnical engineering field, but it did appear and cause catastrophic result to structures and buildings during strong earthquakes. However, due to the rhythmic effect of damaging, a part of the building or adjacent structures may survive or suffer relatively light damage. There is hope and necessity to try to understand the real factors causing such an unusual pattern of ground waving damage.

It is very difficult to run either analytical or numerical assessment to predict such a damage because of the complexity and unknown factors of input seismic waves and seismic response of the site. Nevertheless, it is advisable and worthwhile to make detailed investigation on the conditions of the damaged site aimed at discovering any consistent pattern for future reference and reconstruction.

9.3 COUNTER-MEASURES TO GROUND WAVING DAMAGES

Since theory and methodology have not been fully developed and directed to these problems at present, the counter-measures to be suggested are rather empirical than theoretical.

9.3.1 *Rules of verification*

The following checklists are suggested to help verify ground waving damages ever occurred:

1. Very obvious and unusual manner of damage.
2. Sudden change of local topographic or geological feature which can be identified through site investigation.
3. Existence of higher intensity anomaly. In other words, waving damages are likely to cause local anomaly of higher intensity than that of surrounding areas in a strong earthquake zone.
4. Evidence of permanent waving deformation of the ground.

9.3.2 *Preventive counter measures*

1. In the rehabilitation and reconstruction of an earthquake damaged built-up area, it is reasonably advisable that any project in the same locality where severe damage had occurred, should be either restricted or under strict control of implementing special measures for hazard mitigation. This type of earthquake damages is actually irresistible from the structural point of view. Further, the damage will most likely recur in the same manner in future events, no matter what the source mechanism will be. Therefore, it is suggested to prohibit rebuilding at the same location where total collapse due to ground waving had occurred before.

2. For siting a virgin land in the sense that no previous experience or information about seismic background is available, an experienced earthquake geotechnical engineer may still be able to sort out some doubtful features for the site, which are suspected to be possible factors of ground waving excitation in a strong seismic zone. To follow up, the site conditions should be further studied and compared with those of similar past cases, if any, in other areas.

3. If a very important project is to be built on a site suspected of suffering ground waving, a simple model test of the ground run in the laboratory might be helpful. The test should be run basically under the law of similitude and satisfy the criteria of similarity. Mud may serve as the ground material for simulation. The

topographic and geological features of the site should be incorporated to simulate the boundary conditions. Excitation of seismic input can be performed on a shaking table or even by proper tapping at different frequencies in different directions by trial and error. It should be cautioned that torsional movements at a specific point on the model may still be unclear due to the difficulty of simulating the real earthquake excitation.

[illegible] topographic and geological features of the site should be carefully reported [illegible] boundary conditions [illegible] [illegible] the hazard [illegible] the earthquake [illegible]

References

Adams, J., R.J. Wetmiller, H.S. Hasegawa & J. Drysdale 1991. First surface faulting from a historical intraplate earthquake in North America. *Nature*. Vol. 352.

Ambraseys, N. & J. Jackson 1984. Seismic movements, ground movements and their effects on structures. Surrey Univ. Press.

Boffi, G. & A. Castellani 1988. Effects of surface wave on the rotational components of earthquake motion. *Proceedings of 9th WCEE, Tokyo-Kyoto, Japan*.

ATC-3-06 1978. Tentative provisions for the development of seismic regulations for buildings. Applied Technology Council 3-06 & National Bureau of Standards. USA 1978.

Boffi, G. & A. Castellani 1988. Effects of surface wave on the rotational components of earthquake motion. *Proceedings of 9th WCEE, Tokyo-Kyoto, Japan*.

for new buildings, FEMA 222/Jan. 1992. USA.

Basham, P. & J. Adams 1983. Earthquakes on the continental margin of eastern Canada: Need future large events be confined to the locations of large historical events? Open File Report – United States Geological Survey, No. 83-843, pp. 456–467.

Basham, P.W., D.H. Weichert, F.M. Anglin & M.J. Berry 1982. New probabilistic strong seismic ground motion maps of Canada: A compilation of earthquake source zones, methods and results. *Energy, Mines and Resources Canada*, Earth Physics Branch, Open File 82-33.

Campbell, K.W. 1985. Strong ground motion attenuation relations: A Ten-Year Perspective. *Earthquake Spectra*, Vol. No. 4.

Chavez-Garcia, F.J. & P.-Y. Bard 1989. Effect of random thickness variations on the seismic response of a soft soil layer: Applications to Mexico City. *Proceeding of 4th International Conference on Soil Dynamics & Earthquake Engineering, Mexico City Oct. 1989.*

Committee on Soil Dynamics, Geotechnical Engineering Division, ASCE 1978. Definition of terms related to liquefaction.

Cornell, C.A. 1968. Engineering risk analysis. *Bulletin of the Seismological Society of America*. Vol. 58, No. 5.

Davis, R.O. & J.B. Berrill 1982. Energy dissipation and seismic liquefaction in sands. earthquake engineering and structural dynamics. Vol. 10.

Dobry, R., I.M. Idriss & E. Ng 1969. Duration characteristics of horizontal components of strong-motion earthquake records. *BSSA*. Vol. 68, No. 5.

Dobry, R. et al. 1981. Liquefaction susceptibility from S-wave velocity, in situ testing to evaluate liquefaction susceptibility. *ASCE*, Preprint 81.

Donovan, N.C. & A.M. Becker 1986. Response spectra for building design. *Proc. of Third National Conference on Earthquake Engineering, Charleston, South Carolina*.

References

Adams, J., R.J. Wetmiller, H.S. Hasegawa & J. Drysdale 1991. First surface faulting from a historical intraplate earthquake in North America. *Nature*, Vol. 352.

Ambrasseys, N. & J. Jackson 1984. Seismic movements, ground movements and their effects on structures. Surrey Univ. Press.

Boffi, G. & A. Castellani 1988. Effects of surface wave on the rotational components of earthquake motion. *Proceedings of 9th WCEE, Tokyo-Kyoto, Japan.*

ATC 1978. Tentative provisions for the development of seismic regulations for buildings. Applied Technology Council 3-06 & National Bureau of Standards, USA 1978.

Boffi, G. & A. Castellani 1988. Effects of surface wave on the rotational components of earthquake motion. *Proceedings of 9th WCEE, Tokyo-Kyoto, Japan.*
for new buildings. FEMA 222/Jan, 1992, USA.

Basham, P. & J. Adams 1983. Earthquakes on the continental margin of eastern Canada – Need future large events be confined to the locations of large historical events? Open File Report – United States Geological Survey, No. 83-843, pp. 456-467.

Basham, P.W., D.H. Weichert, F.M. Anglin & M.J. Berry 1982. New probabilistic strong seismic ground motion maps of Canada: A compilation of earthquake source zones, methods and results. *Energy, Mines and Resources Canada*, Earth Physics Branch, Open File 82-33.

Campbell, K.W. 1983. Strong ground motion attenuation relations. *A Ten Year Perspective Earthquake Spectra*, Vol. No.4.

Chavez-Garcia, F.J. & P.-Y. Bard 1989. Effect of random thickness variations on the seismic response of a soft soil layer: Applications to Mexico City. *Proceeding of 4th International Conference on Soil Dynamics & Earthquake Engineering, Mexico City, Oct., 1989.*

Committee on Soil Dynamics, Geotechnical Engineering Division, ASCE 1978. Definition of terms related to liquefaction.

Cornell, C.A. 1968. Engineering risk analysis. *Bulletin of the Seismological Society of America*, Vol. 58, No. 5.

Davis, R.O. & J.B. Berrill 1982. Energy dissipation and seismic liquefaction in sand, earthquake engineering and structural dynamics, Vol. 10.

Dobry, R., I.M. Idriss & E. Ng 1969. Duration characteristics of horizontal components of strong-motion earthquake records. *BSSA*, Vol. 68, No. 5.

Dobry, R. et al. 1981. Liquefaction susceptibility from S wave velocity, in situ testing to evaluate liquefaction susceptibility. *ASCE*, Preprint 81.

Donovan, N.C. & A.M. Becker 1986. Response spectra for building design. *Proc. of Third National Conference on Earthquake Engineering, Charleston, South Carolina.*

Dravinski, M. & M.D. Trifunac 1979. Static, dynamic and rotational components of strong earthquake shaking near faults. University of Southern California, Dept. of Civil Engineering, Los Angeles.

EERI Committee on Seismic Risk 1989. The basics of seismic risk analysis Earthquake Spectra, Vol. 5, No.4.

Eshraghi, H., & M. Dravinski 1990 Amplification of ground motion by three dimensional dipping layers. *Proceeding of 4th US National Conference on Earthquake Engineering, Palm Springs, Ca. May.*

Hakuno, M. & R. Inoue 1975. Effect of soft surface large on the duration time and maximum acceleration of earthquake. *Proceeding of 4th Japan Earthquake Engineering Symposium.*

Hardar, A. & W.H. Tang 1979. Probabilistic evaluation of liquefaction potential. *Journal of Geotechnical Engineering Division, ASCE*, Vol. 105, No. GT2.

Hasegawa, H.S., P.W. Bashaw & M.J. Berry 1981. Attenuation relations for strong seismic ground motion in Canada. *BSSA*, Vol. 71, No.6.

Hoshiba Mitsuyuki, Horike Masaroni, Idei Takafumi & Iwata Tomotaka 1988. Effects of surface geology on seismic motions: Analysis and simulations of seismic motions in a sedimentary basin. *Journal of the Seismological Society of Japan*, Vol. 41, No. 2, Jun.

Husid, R.L. 1969. Analisis de terremoros: Analisis General, Revista del IDIEM, 8, 21-42, Santiago, Chile.

Hu, Y.X. & M.Z. Zhang 1983. Attenuation of ground motion for regions with no ground motion data. *Proceeding of 4th Canadian Conference on Earthquake Engineering.*

Iwasaki, T. 1986. Soil liquefaction studies in Japan, state of the art. *Proceeding of 2nd International Conference on Earthquake Geotechnical Engineering and Soil Dynamics, St. Louis.*

Jin, X. & Z.P. Liao 1991. Engineering prediction of rotational components on ground surface. Earthquake Engineering and Engineering Vibration (in Chinese), Vol. 11, No. 4.

Joyner, W.B. & D.M. Boore 1988. Measurement, characterization and prediction of strong motion. *Proceeding of 2nd International Conference on Earthquake Engineering and Soil Dynamics, St. Louis.*

Kuribayashi, E., T. Iwasaki, Y. Iida & K. Tuji 1973. Effects of seismic and subsoil conditions on earthquake response spectra. *Proc. 5WCEE*, Vol. 1.

Kanai, K. 1961. Observations of micro-tremors. BERI (in Japanese).

Law, K.T. 1991. Analysis of soil liquefaction during the 1988 Saguenay earthquake. *43rd Canadian Geotechnical Conference, Quebec City*, Vol. 1.

Law, K.T., Y.L. Cao & G.N. He 1990. An energy approach for assessing liquefaction potential. *Canadian Geot. Journal*, Vol. 27, No. 3.

Lee, K.L. & Albaisa 1974. Earthquake induced settlement in saturated sands. *Journal of the Geotechnical Engineering Division*, Vol. 100, GT 4.

Lee, T.K. & W. White 1983. *Geotechnical engineering*. Pitman Publishing Inc.

Lee, V.W. & M.D. Trifunac 1987. Rocking strong earthquake acceleration. *EESD*, Vol. 6, No. 2.

Liao, Z.P. & Y. Wei 1989. A semi-empirical method for simulating design accelerograms. *EESD*, Vol. 18, No. 3.

McGuire, R.K. 1977. Seismic design spectra and mapping procedure using hazard analysis based directly on oscillator response. *Earthquake Engineering and Structural Dynamics*, Vol. 5, No. 3.

Niazi, M. 1982. Inferred displacements, velocities and rotations of a long rigid foundation located at El Centro differential array site during the 1979 Imperial Valley, California earthquake. *EESD*, Vol. 14.

Ohori Michihiro & Minami Tadao 1990. *Response analysis of sediment filled valleys due to incident plane waves by two dimensional Aki Larner Method.* Bulletin of the Earthquake Research Institute, Univ. of Tokyo, Vol. 65, Part 4.

Ohsaki, Y., M. Watabe & M. Tohdo 1980. Analysis of seismic ground motion parameters including vertical components. *Proc. 7WCEE*, Vol. 2.

Oliveira, C.S. & B.A. Bolt 1989. Rotational components of surface strong ground motion. *EESD*, Vol. 18, No. 4.

Seed, H.B. 1973. *Stability of earth and rockfill dams during earthquake.* Offprints from Embankment Dam Engineering. John Wiley and Sons.

Seed, H.B. 1979. Soil liquefaction and cyclic mobility evaluation for level ground during earthquake. *J. Geot. Div. ASCE*, Vol. 105 GT2.

Seed, H.B. & P. Dealba 1986. Use of SPT and CPT tests for evaluating the liquefaction resistance of sand. *Proc. of In Situ 86, Special Conference Sponsored by the Geotechnical Engineering Division, ASCE, Blacksburg, Virginia.*

Seed, H.B. & I.M. Idriss 1971. Simplified procedure for evaluating soil liquefaction potential. *Journal of the Soil Mechanics and Foundation Division ASCE*, Vol. 97 No. SM7.

Seed, H.B. & K.L. Lee 1966. Liquefaction of saturated sands during cyclic loading. *Journal of the Soil Mechanics and Foundation Division, ASCE*, Vol. 90 No. SM6.

Shibata, T. & W. Teparaksa 1988. Evaluation of liquefaction potential of soil using performance data. *Journal of Geotechnical Engineering*, Vol. 109, No. 11.

Solonenko, V.P. 1976. *Landslides and collapses in seismic zones and their prediction.* Bulletin of LAEG, No. 15.

Trifunac, M.D. 1982. A note on rotational components of earthquake motions on ground surface for incident body wave. *SDEE*, Vol. 1, No. 1.

Trifunac, M.D. & A.G. Brady 1975. A study on the duration of strong earthquake ground motion. *BSSA*, Vol. 65, No. 3.

Trifunac, M.D. & B. Westermo 1977. A note on the correlation of frequency: Dependent duration of strong earthquake motion with the modified Mercalli intensity and geologic conditions at the recording stations. *BSSA*, Vol. 67, No. 3.

Umemura, H., Y. Ohsaki & M. Watabe 1976. Earthquake parameters for engineering design. *International Conference on the Assessment and Mitigation of Earthquake Risk, Paris.*

Wang, Z.Q. 1981. Macroscopic approach to soil liquefaction. *Proceeding of International Conference on Recent Advances in Geotechnical Earthquake Engineering and Soil Dynamics, St. Louis*, Vol. 1.

Wang, Z.Q. 1983. Surface displacement in relation to shallow surface fractures and deep faulting. *Canadian Geotechnical Journal*, Vol. 20, No. 1.

Wang, Z.Q. et al. 1986. *Testing techniques in geotechnical engineering* (in Chinese). Building Industry Press, China.

Wang, Z.Q. & M. Wang 1989. Prediction of soil liquefaction damage on natural ground. *Special Vol. TC 12th ICSMFE, Rio, Brazil.*

Wiegel, R.L. 1970. Earthquake engineering, Prentice-Hall, Inc.

Yudemen, M.S. 1980. Uncertainty analysis in the evaluation of seismic risk. *Proceeding of the Seventh World Conference on Earthquake Engineering*, Vol. 1.

REFERENCES IN CHINESE

[1] 盧榮儉 (1979)："地震烈度與地面運動的關係"，中國科學院工程力學研究所報告

[2] 渡部丹、山内泰之、大川出、千葉修、藤堂正喜 (1984)："地面運動參數的特征"，國際地震工程專題討論會，同濟大學出版社

[3] 胡聿賢 (1988)：地震工程學，地震出版社

[4] 岡本舜三 (1971)：抗震工程學，中譯本，中國建築工業出版社

[5] 胡聿賢等 (1982)："基岩地震動參數與震能和距離的關係"，地震學報，4卷，2期

[6] 胡聿賢、時振樑等 (1988)：重要工程中的地震問題，地震出版社

[7] 伊德里斯、I.M.等著，謝君斐等譯 (1985)：地震工程和土動力問題譯文集，地震出版社

[8] 冶金工業部建築研究總院工程抗震研究室 (1982)：九國抗震設計規范匯編，地震出版社

[9] 章在墉 (1979)："地面運動持續時間的研究現狀和前景"，中國科學院工程力學研究所報告，79 -- 014

[10] 謝毓壽 (1988)：地震烈度，地震出版社

[11] 黃偉玉、常向東等 (1989)："地震活動時空不均勻性年平均發生率估計"，中國地震，5,2 期

[12] 廖振鵬、田啟文、孫平善等 (1985)："四川省大渡河瀑布水電站壩址地震危險性分析報告"，國家地震局工程力學研究所報告，85 -- 062

[13] 王鍾琦、朱小林、孫廣忠、唐家洪、劉雙光、黃世銘 (1986)：岩土工程測試技術，第五章，中國建築工業出版社，1988年第二版

[14] 廖振鵬、田啟文、孫平善 (1986)："大連市地震小區劃的地震輸入"，中國抗震防災論文集，(上冊)

[15] 王阜 (1986)："地震發生的隨機模型"，世界地震工程，第二期

[16] 廖振鵬主編 (1989)：地震小區劃理論與實踐，地震出版社

[17] A. Der Kiureghian, A. H-S. Ang (洪華生)："地震危險性分析用的斷裂 -- 破裂模型"，國外地震工程，4期，1981，譯自 BSSA, Vol.67, No.4, 1977

[18] 國家核安全局、國家地震局 (1987)：安全導則 "核電廠廠址選擇中的地震問題"

[19] 左惠強 (1988)："地震活動性參數在地震危險性分析中的作用"，國家地震局工程力學研究所

[20] 蔣溥 (1987)："場地地震效應及其預測"，國家地震局地質所

[21] 田啟文，廖振鵬，孫平善 (1986)，根據烈度資料估算我國地震動參數衰減規律，地震工程与工程振動，第六卷第一期。

[22] 胡聿賢 (1986): "參考唐山地震確定的地震動衰減規律", 土木工程學報, 第19卷第一期

[23] 王廣軍, 蘇經宇 (1985): "國外抗震設計規范中設計反應譜的應用及進展", 建築結构, 5期

[24] 周錫元, 王廣軍, 蘇經宇 (1990): "場地、地基、設計地震", 地震出版社, 1990年

[25] 章在墉 (1984): "地震危險性分析概論", 同濟大學結构理論分析研究所

[26] 渡部丹, 山内泰之等 (1984): "地震地面運動參數特征", 國際地震工程專題討論會譯文集, 同濟大學出版社

[27] 嚴建武, 吳洪剛, 張永凱 (1988): "黃河小浪底水庫壩址區地震動參數的研究", 工程抗震, 1期

[28] 廖振鵬, 李大華 (1989): "設計地震反應譜的雙參數標定模型" 地震小區划理論与實踐, 地震出版社, 1989年

[29] 田啟文等 (1987); "呼和浩特市地震危險性分析", 國家地震局工程力學研究所

[30] 胡聿賢, 何訓 (1989): "考慮相位譜的人造地震動反應譜擬合", 地震小區划理論与實踐, 地震出版社, 1989年

[31] 廖振鵬, 魏穎 (1989): "設計地震加速度圖的合成", 地震小區划理論与實踐, 地震出版社, 1989年

[32] 田治米辰雄等 (王興健等譯, 1982): 地震与震害, 地震出版社, 1982

[33] 周錫元, 王廣軍, 蘇經宇 (1989): "規范用抗震設計反應譜的修訂趨向", 地震小區划理論与實踐, 地震出版社, 1989年

[34] 岩土工程勘察規范編制組: 岩土工程勘察規范 (1989年草案)

[35] 王鍾琦, 謝君斐, 石兆吉 (1983): 地震工程地質導論, 地震出版社, 1988年

[36] J.B.伯里爾 (1988): 跨越斷層建築物的危險估計方法, 地震危險性評定与地震區划, 地震出版社, (中譯本)

[37] 方鴻琪、王鍾琦等 (1981): 唐山地震區地震工程地質研究, 中國建築科學研究院勘察技術所

[38] 劉穎, 謝君斐等 (1984): 砂土振動液化, 地震出版社1984

[39] 中華人民共和國國家標准: 建築抗震設計規范 (GBJH-89), 中國建築工業出版社, 1989年

[40] 石兆吉 (1986): "判別水平土層液化勢的剪切波速法", 水文地質与工程地質, VOI.4

[41] 汪敏 (1990): "地基液化勢新評價方法", 工程勘察, 1990年第1期

[42] 李天池 (1979): "地震与滑坡的關系及地震滑坡預測的探討", 滑坡文集 (第二集), 人民鐵道出版社, 1979年

[43] 中華人民共和國水利電力部: 水工建築抗震設計規范 (SDJ10-78) 水利電力出版社, 1979年